'A beautifully written exploration of perhaps the most important question in science. I felt I was being given rare access to a truly deep understanding of a complex and profound subject. This is the best introduction to modern biology I've read.'

Brian Cox

'In this vibrant, lively book, Sir Paul Nurse, discoverer of some of the crucial genes that control the division of cells, takes a deep dive into biology by illuminating five of the essential characteristics of "life". The writing is so spirited and knowledgeable – and the five sections so full of wondrous revelations – that I could not put it down. This is a book that will inspire a generation of biologists.'

Siddhartha Mukherjee

'A masterful overview of biology that draws together big ideas, luminous details and personal insights. You emerge with a more profound sense of wonder about the diversity, complexity and interconnectedness of living organisms. It's the biggest question in biology. And this book represents the best answer I've ever seen. Paul Nurse is a rare life-form – a Nobel-winning scientist and a brilliant communicator.'

Alice Roberts

'Paul Nurse is about as distinguished a scientist as there could be. He is also a great communicator. This book explains, in a way that is both clear and elegant, how the processes of life unfold, and does as much as science can to answer the question posed by the title. It's also profoundly important, at a time when the world is connected so closely that any new illness can sweep from nation to nation with immense speed, that all of us – including politicians – should be as well-informed as possible. This book provides the sort of clarity and understanding that could save many thousands of lives. I learned a great deal, and I enjoyed the process enormously.'

Philip Pullman

'Paul Nurse provides a concise, lucid response to an age-old question. His writing is not just informed by long experience, but also wise, visionary and personal. I read the book in one sitting, and felt exhilarated by the end, as though I'd run for miles – from the author's own garden into the interior of the cell, back in time to humankind's most distant ancestors, and through the laboratory of a dedicated scientist at work on what he most loves to do.'

Dava Sobel

WHAT IS LIFE?

www.**davidficklingbooks**.com

FROM NOBEL
PRIZE WINNER

PAUL
NURSE

WHAT IS
LIFE?

EDITED BY
BEN MARTNOGA

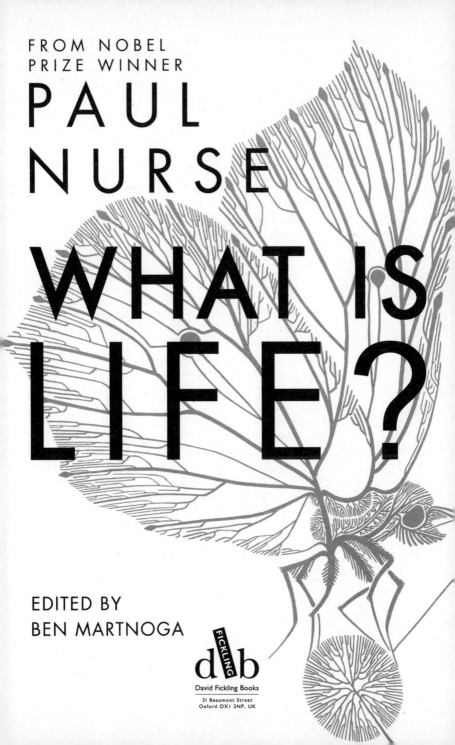

db
David Fickling Books

31 Beaumont Street
Oxford OX1 2NP, UK

What is Life?
is a
DAVID FICKLING BOOK

First Published in hardback in Great Britain in 2020 by
David Fickling Books,
31 Beaumont Street,
Oxford, OX1 2NP
This edition published in 2021

Edited by Ben Martynoga

Cover design by Paul Duffield

978-1-78845-142-0

3 5 7 9 10 8 6 4 2

DAVID FICKLING BOOKS Reg. No. 8340307

A CIP catalogue record for this book is available from the British Library.

Typeset in 12/18pt Goudy Old Style by Falcon Oast Graphic Art Ltd.
Printed and bound in Great Britain by Clays Ltd, Elcograf S.p.A.

Extract from *The Temple of Nature* by Erasmus Darwin
published by permission of Timaios Press.

To Andy Martynoga (Yog), friend and father
and my grandchildren:
Zoe, Joseph, Owen and Joshua
and their generation
who will need to care for Life on our planet

CONTENTS

INTRODUCTION

It may have been a butterfly that first started me thinking seriously about biology. It was early spring; I was perhaps twelve or thirteen years old and sitting in the garden when a quivering yellow butterfly flew over the fence. It turned, hovered and briefly settled – just long enough for me to notice the elaborate veins and spots on its wings. Then a shadow disturbed it and it took flight again, disappearing over the opposite fence. That intricate, perfectly formed butterfly made me think. It was both utterly different to me and yet somehow familiar too. Like me, it was so obviously alive: it could move, it could sense, it could respond, it seemed so full of *purpose*. I found myself wondering:

what does it really mean to be alive? In short, what is life?

I have been thinking about this question for much of my life, but finding a satisfactory answer is not easy. Perhaps surprisingly, there is no standard definition of life, although scientists have wrestled with this question across the ages. Even the title of this book, *What is Life?*, has been shamelessly stolen from a physicist, Erwin Schrödinger, who published an influential book of the same name in 1944. His main focus was on one important aspect of life: how living things maintained such impressive order and uniformity for generation after generation in a universe that is, according to the Second Law of Thermodynamics, constantly moving towards a state of disorder and chaos. Schrödinger quite rightly saw this as a big question, and he believed that understanding inheritance – that is what genes are and how they are passed on faithfully between generations – was key.

In this book I ask the same question – What is life? – but I do not think that *only* deciphering inheritance will give us a complete answer. Instead

I will consider five of biology's great ideas, using them as steps that we can climb, one at a time, to get a clearer view of how life works. These ideas have mostly been around for some time, and are generally well accepted for explaining how living organisms function. But I will draw these different ideas together in new ways, and use them to develop a set of unifying principles that define life. Hopefully they will help you see the living world through fresh eyes.

I should say, right at the start, that we biologists often shy away from talking about great ideas and grand theories. In this respect we are rather different from physicists. We sometimes give the impression that we are more comfortable immersing ourselves in details, catalogues and descriptions, whether that's listing all the species in a particular habitat, counting the hairs on a beetle's leg, or sequencing thousands of genes. Perhaps it is nature's bewildering, even overwhelming, diversity that makes it seem hard to seek out simple theories and unifying ideas. But important overarching ideas of this kind *do* exist in biology, and they help us make sense of life in all its complexity.

The five ideas I will explain to you are: 'The Cell', 'The Gene', 'Evolution by Natural Selection', 'Life as Chemistry' and 'Life as Information'. As well as explaining where they came from, why they are important, and how they interact, I want to show you that they are still changing and being further developed today, as scientists all over the world make new discoveries. I also want to give you a taste of what it's like to be engaged in scientific discovery, so I will introduce you to the scientists who made these advances, some of whom I knew personally. I will also tell you stories of my own experiences of doing research in the laboratory, the 'lab', including the hunches, the frustrations, the luck and the rare but wonderful moments of genuinely new insight. My aim is for you to share in the thrill of scientific discovery and to experience the satisfaction that comes through a growing understanding of the natural world.

Human activity is pushing our climate and many of the ecosystems it supports to the edges of – or even beyond – what they can bear. To maintain life as we know it, we are going to need all the insights we can get from studying the living world. That is why in the years and decades ahead, biology will increasingly steer the choices we make about how people live, are born, fed, healed and protected from pandemics. I will describe some of the applications of biological knowledge and the difficult trade-offs, ethical uncertainties and the possible unintended consequences that they can give rise to. But before we can join the growing debates that surround these topics, we first need to ask what life is and how it functions.

We live in a vast and awe-inspiring universe, but the life that thrives right here in our tiny corner of that greater whole is one of its most fascinating and mysterious parts. The five ideas in this book will act like steps that we will move up, progressively revealing principles that define life on Earth. This will also help us think about how life on our planet might have first got started and what life might be like should we ever encounter it elsewhere in the universe. Whatever your starting point – even if you think that you know little or nothing about science – by the time you have finished this book, my goal is for you to have a better sense of how you, me, that delicate yellow butterfly and all other living things on our planet are connected.

It is my hope that, together, we will be closer to understanding what life is.

1. THE CELL

Biology's Atom

I saw my first cell when I was at school, not long after my encounter with the yellow butterfly. My class had germinated onion seedlings and squashed their roots under a microscope slide to see what they were made from. My inspirational biology teacher, Keith Neal, explained that we would see cells, the basic unit of life. And there they were: neat arrays of box-like cells, all stacked up in orderly columns. How impressive it seemed that the growth and division of those tiny cells were enough to push the roots of an onion down through the soil, to provide the growing plant with water, nutrients and anchorage.

As I learned more about cells, my sense of

wonder only grew. Cells come in an incredible variety of shapes and sizes. Most of them are too small to be seen with the naked eye – they are truly minute. Individual cells of a type of parasitic bacteria that can infect the bladder could line up 3,000-abreast across a one-millimetre gap. Other cells are immense. If you had an egg for breakfast, consider the fact that the whole of its yolk is just one single cell. Some cells in our bodies are also huge. There are, for example, individual nerve cells that reach from the base of your spine all the way to the tip of your big toe. That means those cells can each be about a metre long!

Startling as all this diversity is, what is most interesting for me is what all cells have in common. Scientists are always interested in identifying fundamental units, the best example being the atom as the basic unit of matter. Biology's atom is the cell. Cells are not only the basic structural unit of all living organisms, they are also the basic functional unit of life. What I mean by this is that cells are the smallest entities that have the core characteristics of life. This is the basis of what biologists call *cell theory*: to the best of our knowledge, everything that is alive on the planet is either a cell or made from a collection of cells. The cell is the simplest thing that can be said, definitively, to be alive.

Cell theory is about a century and a half old, and it has become one of biology's crucial foundations. Given the importance of this idea for understanding biology, I find it surprising that it has not caught the public imagination more than it has. This might be because most people are taught in school biology classes to think of cells as mere building blocks for more complex beings, when the reality is much more interesting.

The story of the cell begins in 1665 with Robert Hooke, a member of the newly formed Royal Society of London, one of the first science academies in the world. As is so often the case in science, it was a new technology that triggered his discovery. Since most cells are too small to see with the naked eye, their discovery had to wait for the invention of the microscope in the early seventeenth century. Scientists are often a combination of theorist and skilled artisan, and this was certainly true of Hooke, who was equally comfortable exploring the frontiers of physics, architecture or biology as he was inventing scientific instruments. He built his own microscopes, which he then used to explore the strange worlds hidden beyond the reach of the naked eye.

One of the things Hooke looked at was a thin slice of cork. He saw that the cork wood was made up of row after row of walled cavities, very similar to the cells in the onion root tips I saw as a schoolboy 300 years later. Hooke named these cells after the Latin word *cella*, meaning a small room or cubicle. At that time Hooke did not know that the cells he had drawn were in fact not only the basic component of all plants, but of all life.

Not long after Hooke, the Dutch researcher Anton van Leeuwenhoek made another crucial observation when he discovered single-celled life. He spotted these microscopic organisms swimming in samples of pond water and growing in the plaque he scraped from his teeth: an observation that disturbed him, since he was rather proud of his dental hygiene! He gave these tiny beings an endearing name, that we no longer use today, 'animalcules'. Those he found flourishing between his teeth were, in fact, the first bacteria ever described. Leeuwenhoek had stumbled across an entire new domain of minute single-celled life forms.

We now know that bacteria and other sorts of microbial cells ('microbe' is a general term for all microscopic organisms that can live as single cells) are by far the most numerous life forms on Earth. They inhabit every environment, from the high atmosphere to the depths of the Earth's crust. Without them, life would come to a standstill. They break down waste, build soils, recycle nutrients and capture from the air the nitrogen that plants and animals need to grow. And when scientists look at our own bodies, they see that for each and every one of our 30 trillion or more human cells, we have at least one microbial cell. You – and every other human being – are not an isolated, individual entity, but a huge and constantly changing colony made up of human and non-human cells. These cells of microscopic bacteria and fungi live *on* us and *in* us, affecting how we digest food and fight illnesses.

But before the seventeenth century, nobody had any idea that these invisible cells even existed, let alone that they worked according to the same basic principles as all other more visible life forms.

During the eighteenth century and into the beginning of the nineteenth century, micro-scopes and microscopic techniques improved, and very soon scientists were identifying cells from all manner of different creatures. Some began to speculate that all plants and animals were built from collections of those animalcules that Leeuwenhoek had identified several generations before them. Then, after a long gestation, the cell theory was finally fully born. In 1839 the bota-nist Matthias Schleiden and zoologist Theodore Schwann, summarized work from themselves and many other researchers, and wrote 'we have seen that all organisms are composed of essentially like parts, namely of cells'. Science had reached the illuminating conclusion that the cell is the funda-mental structural unit of life.

The implications of this insight deepened further when biologists realized that every cell is a life form in its own right. This idea was captured by the pioneering pathologist Rudolf Virchow, when he wrote in 1858 'that every animal appears as a sum of vital units, each of which bears in itself the complete characteristics of life'.

What this means is that all cells are themselves alive. Biologists demonstrate this most vividly when they take cells from the multicellular bodies of animals or plants and keep them alive in glass or plastic vessels, often flat-bottomed vessels called Petri dishes. Some of these cell lines have been growing in laboratories around the world for decades on end. They let researchers study biological processes without needing to deal with the complexity of whole organisms. Cells are active; they can move and respond to the environment, and their contents are always in motion. Compared to a whole organism, like an animal or a plant, a cell may seem simple, but it is definitely alive.

There was, however, an important gap in cell theory, as originally formulated by Schleiden and

Schwann. It did not describe how new cells came into being. That gap closed when biologists recognized that cells reproduce by dividing themselves from one cell into two, and concluded that cells are only ever made by the division of a pre-existing cell in two. Virchow popularized this idea with a Latin epigram: '*Omnis cellula e cellula*', that is, all cells come from cells. This phrase also helped to counter the incorrect idea, still popular amongst some at the time, that life arises spontaneously from inert matter all the time – it does not.

Cell division is the basis of the growth and development of all living organisms. It is the first critical step in the transformation of a single, uniform fertilized egg of an animal into a ball of cells and then, eventually, into a highly complex and organized living being, an embryo. It all begins with a cell dividing and producing two cells which can take on different identities. The entire development of the embryo that then takes place is based on this same process – repeated rounds of cell division, followed by the creation of an ever more elaborately patterned embryo, as cells mature into increasingly specialized tissues and organs. This means that all living organisms, regardless of their size or complexity, emerge from a single cell. I think we would all respect cells a little more if we remembered that every one of us was once a single cell, formed when a sperm and an egg fused at the moment of our conception.

Cell division also explains the apparently miraculous ways the body heals itself. If you were to cut yourself with the edge of this page, it would be localized cell division around the cut that would repair the wound, helping to maintain a healthy body. Cancers, however, are the unfortunate counterpoint to the body's ability to instigate new rounds of cell division. Cancer is caused by the uncontrolled growth and division of cells that can spread their malignancy, damaging or even killing the body.

Growth, repair, degeneration and malignancy are all linked to changes in the properties of our cells, in sickness and in health, in youth and in old age. In fact, most diseases can be traced back to the malfunction of cells, and understanding what goes wrong in cells underpins how we develop new ways to treat disease.

Cell theory continues to influence the trajectory of research in the life sciences and in medical practice. It drastically shaped the course of my life too. Ever since my thirteen-year-old self squinted down a microscope and saw the cells of that onion root tip, I have been curious about cells and how they work. When I started as a biology researcher, I decided to study cells, in particular how cells reproduce themselves and control their division.

The cells I started to work with in the 1970s were yeast cells, which most people think are only good for making wine, beer or bread, not for tackling fundamental biological problems. But they are, in fact, a great model for understanding how cells of more complex organisms work. Yeast is a fungus, but its cells are surprisingly similar to plant and animal cells. They are also small, relatively simple, and grow quickly and inexpensively when fed on simple nutrients. In the lab we grow them either floating freely in a liquid broth or on top of a layer of jelly in a plastic Petri dish, where they form cream-coloured colonies a few millimetres across, each containing many millions of cells. Despite, or more accurately, because of their simplicity, yeast cells have helped us to understand how cells divide in most living organisms, including human cells. Quite a lot of what we know about the uncontrolled cell divisions of cancer cells came first from studying the humble yeasts.

Cells are the basic unit of life. They are individual living entities, surrounded by membranes made from fat-like lipids. But, just as atoms contain electrons and protons, cells contain smaller components too. Microscopes today are very powerful and biologists use them to reveal the intricate and often very beautiful structures within cells. The largest of these structures are called *organelles*, which are each wrapped in their own layer of membrane. Of these, the *nucleus* is a command centre of the cell, since it contains the genetic instructions written into the chromosomes, while *mitochondria* – and there can be hundreds of these in some cells – act as miniature power plants, supplying the cell with the energy it needs to grow and survive. A variety of other containers and compartments within cells perform sophisticated logistics functions, building, breaking down or recycling cellular parts, as well as shuttling materials in and out of the cell and transporting them around the cell's interior.

Not all living organisms are based on cells that contain these membrane-bounded organelles and complex internal structures, however. The presence or absence of a nucleus divides life into two major branches. Those organisms whose cells contain a nucleus – such as animals, plants and fungi – are called *eukaryotes*. Those without a nucleus are called *prokaryotes*, which are either bacteria or archaea. Archaea appear to be similar to bacteria in terms of their size and structure but are actually their distant relatives. In some respects their molecular workings are more similar to eukaryotes like us, than they are to bacteria.

A critically important part of a cell, be it a prokaryote or a eukaryote, is its outer membrane. Although just two molecules thick, this outer membrane forms a flexible 'wall' or barrier that separates each cell from its environment, defining what is 'in' and what is 'out'. Both philosophically and practically, this barrier is crucial. Ultimately, it explains why life forms can successfully resist the overall drive of the universe towards disorder and chaos. Within their insulating membranes, cells can establish and cultivate the order they need to operate, whilst at the same time creating disorder in their local surroundings outside the cell. That way life does not contravene the Second Law of Thermodynamics.

All cells can detect and respond to changes in their inner state and in the state of the world around them. So although separated from the environment they live in, they are in close communication with their surroundings. They are also constantly active and working to maintain the internal conditions that allow them to survive and to flourish. They share this with more visible living organisms, such as the butterfly I watched as a child, or for that matter with ourselves.

In fact, cells share many characteristics with all kinds of animals, plants and fungi. They grow, they reproduce, they maintain themselves, and in doing all of this they display a sense of purpose: an imperative to persist, to stay alive and to reproduce, come what may. All cells, from the bacteria Leeuwenhoek found between his teeth to the neurons that allow you to read these words, share these properties with all living beings. Understanding how cells work brings us closer to understanding how life works.

Core to the existence of the cell are the genes, which we will turn to next. These encode instructions that each cell uses to build and organize itself, and they must be passed to every new generation when cells and organisms reproduce.

2. THE GENE

The Test of Time

I have two daughters and four grandchildren. All of them are wonderfully unique. For example, one of my daughters, Sarah, is a TV producer and the other, Emily, a professor of physics. But there are also characteristics that they share between themselves, their children, with me, and my wife Anne. Family resemblances can be strong or subtle – height, eye colour, the curve of the mouth or nose, even particular mannerisms or facial expressions. There are many variations too, but the continuity between generations is undeniably there.

The existence of similarities between parents and offspring is a defining characteristic of all living organisms. It was something that Aristotle and other classical thinkers recognized long ago, but the basis of biological inheritance remained a stubborn mystery. Various explanations were given over the years, some of which sound a bit peculiar today. Aristotle, for example, suspected that mothers only influenced the development of their unborn children in the same way that the qualities of a particular soil influenced the growth of a plant from a seed. Others thought that the explanation was 'blending of the blood'; that is the idea that children inherit an average mixture of their two parents' characteristics.

It took the discovery of the gene to pave the way to a more realistic understanding of how inheritance works. As well as providing a way to help make sense of the complicated mixture of resemblances and unique characteristics that run through families, genes are the key source of information life uses to build, maintain and reproduce cells and, by extension, organisms made from cells.

Gregor Mendel, Abbot of Brno Monastery, now in the Czech Republic, was the first person to make some sense of the mysteries of inheritance.

But he didn't do this by studying the often baffling patterns of inheritance in human families. Instead, he carried out careful experiments with pea plants, hatching the ideas that led eventually to the discovery of the things we now call genes.

Mendel wasn't the first person to use scientific experiments to ask questions about inheritance, nor even the first to use plants to look for answers. These earlier plant breeders had described how some characteristics of plants were passed through the generations in counterintuitive ways. The offspring of a cross between two different parental plants would sometimes look like a blend between the two of them. For example, crossing a purple-flowered plant and a white-flowered plant might give rise to a pink-flowered plant. But other characteristics always seemed to dominate in a particular generation. In these situations, the offspring of a purple-flowered plant and a white-flowered plant would all have purple flowers, for example. The early pioneers had gathered lots of intriguing clues, but none of them had managed to reach a satisfying understanding of how genetic inheritance operates in plants, let alone explain how it works in essentially all living things, including us humans. That, however, is exactly what Mendel started to reveal with his work on peas.

In 1981, in the middle of the Cold War, I went on my own pilgrimage to the Augustinian monastery in Brno to see where Mendel worked. This was long before it had become the tourist attraction that it is today. The garden, then rather overgrown, was surprisingly big. I could easily imagine the rows upon rows of peas that Mendel once grew there. He had studied physical sciences at the University of Vienna but failed to qualify as a teacher. However, something from his training in physics stuck with him. He clearly understood that he would need a lot of data: big samples are more likely to uncover important patterns. Some of his experiments involved 10,000 different pea plants. None of the plant breeders before him had taken such a rigorous, extensive quantitative approach.

To reduce the complexity of his experiments, Mendel focused only on characteristics that displayed clear-cut differences. Over several years, he carefully recorded the outcomes of the crosses he set up, and detected patterns that others had missed. Most importantly, he observed distinctive arithmetic ratios of pea plants that either exhibited or lacked specific characteristics, such as particular flower colours or seed shapes. One of the crucial things Mendel did was to describe these ratios in terms of a mathematical series. This led him to propose that the male pollen and female ovules within the pea flowers, contain things he called 'elements' that are associated with the different characteristics of the parental plants. When these elements come together through fertilization they influence the characteristics of the next generation of plants. However, Mendel did not know what the elements were or how they might work.

By intriguing coincidence, another famous biologist, Charles Darwin, was studying plant crosses with flowers called snapdragons at around the same time as Mendel. He observed similar ratios, but did not try and interpret what they might mean. In any case, Mendel's work was almost entirely ignored by his contemporaries and it was another whole generation before anyone took him seriously.

Then, around 1900, other biologists working independently repeated Mendel's results, developed them further, and started making more explicit predictions about *how* inheritance might work. This work led to the theory of Mendelism, named in the pioneering monk's honour, and the birth of genetics. Now the world began to take notice.

Mendelism proposes that inherited characteristics are determined by the presence of physical particles, which exist as a pair. These 'particles' are what Mendel called 'elements' and we now call genes. Mendelism did not have much to say about what these particles were, but it described in a very precise way how they are inherited. And most crucially, it gradually became clear that these conclusions not only applied to peas, but to all sexually reproducing species, from a yeast to humans and all organisms in between. Every one of your genes exists as a pair; you inherited one from each of your biological parents. They were transmitted through the sperm and egg that fused together at the moment of your conception.

Science had not stood still during the final third of the nineteenth century when Mendel's discoveries had lain fallow. In particular, research-ers had finally managed to get a clearer view of cells engaged in the process of cell division. When these observations were eventually linked to the inherited particles proposed by Mendelism, the gene's central role in life came into a sharper focus.

An early clue was the discovery of microscopic structures within cells that looked like tiny threads. These were first spotted in the 1870s by a German military physician turned cell biologist called Walther Flemming. Using the best microscopes of his day, he described how these microscopic threads behaved in an intriguing way. As a cell got ready to divide, Flemming saw these threads splitting in half lengthwise, before becoming shorter and thicker. Then, when the cell divided into two, the threads were *separated*, with one half ending up in each of the newly formed daugh-ter cells.

What Flemming was looking at, but did not fully understand at the time, was the physical manifestation of the genes, the inheritable particles proposed by Mendelism. What Flemming called 'threads', we now call *chromosomes*. Chromosomes are the physical structures present in all cells that contain the genes.

Around the same time, another crucial clue about genes and chromosomes emerged from an unlikely source: the fertilized eggs of parasitic roundworms. When the Belgian biologist Edouard van Beneden examined carefully the very earliest stages of roundworm development, he saw through his microscope that the first cell of each newly fertilized embryo contained four chromosomes. It received precisely two from the egg and two from the sperm.

This fitted exactly with the predictions of Mendelism – two sets of paired genes, brought together at the moment of fertilization. Van Beneden's results have since been confirmed many times over. There are half the chromosomes in eggs and sperm, and the full number of chromosomes are formed when the two fuse to make a fertilized egg. We now know that what is true for sexual reproduction in roundworms is true for all eukaryotic life, including us humans.

The number of chromosomes varies widely: pea plants have 14 in each cell, we have 46, and the cells of the Atlas blue butterfly have more than 400. Fortunately for van Beneden, the roundworm has just four. If there had been more chromosomes, he could not have easily counted them. By paying close attention to the relatively simple case of the roundworm, van Beneden glimpsed a universal truth about genetic inheritance. Starting with a clearly interpretable experiment with a simple biological system can lead to a wider insight relevant more generally to how life works. For precisely this reason I have spent most of my career investigating the simple and easily studied yeast cells, rather than more complex human cells.

Putting the discoveries made by Flemming and van Beneden together, it became clear that chromosomes convey genes both between the generations of dividing cells, and also between the generations of whole organisms. Apart from a few specialized exceptions – like red blood cells which, as they mature, lose their entire nucleus and therefore all their genes – every cell in your body contains a copy of your entire complement of genes. Together, those genes play a big role in directing the development of a fully formed body from a lone fertilized egg cell. And across the entire lifespan of each living organism, the genes provide each cell with essential information it needs to build and maintain itself. It follows, therefore, that every time a cell divides, the entire set of genes must be copied and shared equally between the two newly formed cells. This means that cell division is the fundamental example of reproduction in biology.

The next great challenge for biologists was to understand what genes actually are and how they work. The first big insight came in 1944, when a small group of scientists in New York, led by the microbiologist Oswald Avery, carried out an experiment that identified the substance that genes are made from. Avery and his colleagues were studying bacteria that cause pneumonia. They knew that harmless strains of these bacteria could be transformed into dangerous, virulent forms when they were mixed with remnants of dead cells from a virulent strain. Critically, this change was inheritable; once they became virulent, the bacteria passed that characteristic on to all their descendants. This led Avery to reason that a gene or genes had been passed, as a chemical entity, from the remains of the dead, harmful bacteria to the live, harmless bacteria, changing their nature for ever. He realized that if he could find the part of the dead bacteria responsible for this *genetic* transformation, he could finally show the world what genes are made of.

It turned out that it was, in fact, a substance called deoxyribonucleic acid – which you will probably recognize from its more famous acronym DNA – that had the key transformative property. By then, it was widely known that the gene-carrying chromosomes within cells contained DNA, but most biologists thought that DNA was too simple and boring a molecule to be responsible for such a complex phenomenon as heredity. They were wrong.

Each of your chromosomes has at its core a single, unbroken molecule of DNA. These can be extremely long and each can contain hundreds or even thousands of genes arranged in a chain, one after another. Human chromosome number 2, for example, contains a string of over 1,300 different genes, and if you stretched that piece of DNA out, it would measure more than 8 cm in length. This leads to the extraordinary statistic that, together, the 46 chromosomes in each of your tiny cells would add up to more than two metres of DNA. Through some miracle of packing, it all fits into a cell that measures no more than a few thousandths of a millimetre across. What is more, if you could somehow join together and then stretch out all the DNA coiled up inside your body's several trillion cells into a single, slender thread, it'd be about 20 billion kilometres long. That's long enough to stretch from Earth to the sun and back sixty-five times!

Avery was quite a modest man and didn't make much of a fanfare about his discovery, while some biologists were critical of his conclusion. But he was right: genes are made of DNA. Once that truth finally sank in, it signalled the birth of a new era for genetics and for biology as a whole. Genes could finally be understood as chemical entities: stable collections of atoms that obeyed the laws of physics and chemistry.

However, it was the elucidation of the structure of DNA, in 1953, that truly ushered in this brave new era. Most of the important discoveries in biology depend upon the work of many scientists who, over years or decades, have scratched away at the nature of reality to gradually reveal an important truth. But sometimes spectacular insights are achieved much more quickly. So it was with the structure of DNA. In a matter of months, three scientists working in London – Rosalind Franklin, Raymond Gosling and Maurice Wilkins – did the crucial experiments, and then Francis Crick and James Watson, based in Cambridge, interpreted the experimental data and correctly deduced the structure of DNA. Furthermore, they quickly grasped what it meant for life.

Later, when they were older, I got to know

both Crick and Watson quite well. They made a contrasting pair. Francis Crick had a razor-sharp, logically incisive mind. He'd slice problems up until they literally melted under his gaze. James Watson had a brilliant intuition, jumping to conclusions others had not seen, although it was not always clear how he got there. Both were confident and outspoken, and although sometimes critical, they were also highly interactive with young scientists. Together, they were a formidable combination.

The real beauty of the DNA double helix they proposed is not the elegance of the gracefully spiralling structure itself. Rather, it is the way the structure explains the two key things that the hereditary material must do to underpin the survival and perpetuation of life. First, DNA must encode the information that cells and whole organisms need to grow, maintain and reproduce themselves. Second, it must be able to replicate itself, precisely and reliably, so that each new cell, and each new organism, can inherit a complete set of genetic instructions.

DNA's helical structure, which you can think of as a twisted ladder, explains both of these critical functions. Let's look at how DNA carries information. The rungs of the ladder are each made from links that form between pairs of chemical molecules called nucleotide bases. These bases come in just four different types, which we can abbreviate from Adenine, Thymine, Guanine and Cytosine, to A, T, G and C. The order in which these four bases appear along each of the two rails, or strands, of the DNA ladder functions as an information-containing code. This is just like the meaning that is communicated by the ordered string of letters that makes up this sentence that you are reading. Each gene is a defined stretch of this DNA code that contains a message for the cell. That message might be the instruction to produce a pigment that will determine the colour of a person's eyes, make the cells of a pea flower purple, or make a pneumonia bacterium more virulent, for example. The cell obtains these messages from DNA by 'reading' this genetic code and putting that information to work.

Then there's the need to make accurate copies of DNA, so all the information in the genes can be passed faithfully from one generation of cells or organisms to the next. The shape and chemical properties of the two nucleotide bases that make up each rung of the ladder ensure that the bases can only pair up in a single, precise way. A can only pair with T, and G can only pair with C. This means that if you know the order of bases along one strand of DNA, you immediately know the order of the nucleotide bases on the other strand. It follows, therefore, that if you break the double helix apart into its two strands, each strand can act as a template to recreate a perfect copy of its original partner strand. As soon as Crick and Watson saw that DNA was built this way, they knew that this must be the way that cells copy the DNA making up their chromosomes, and with it their genes.

Genes exert their major influence on the behaviour of cells, and ultimately whole organisms, by instructing the cell how to construct particular proteins. This information is central to life because proteins are the things that do most of the work in the cell – most of the cell's enzymes, structures and operational systems are made from proteins. To do this, cells translate between two alphabets: the four-letter alphabet of DNA, made up of the 'letters' A, T, G and C; and the more complex alphabet of proteins, which consists of ordered strings of 20 different building blocks called amino acids. By the early 1960s this basic relationship between genes and proteins was understood, but nobody knew how the cell translated information from the language of DNA into the language of proteins.

This relationship is known as the 'genetic code' and it presented biologists with a true cryptographic puzzle. The code was finally cracked during the late 1960s and early 1970s, by a succession of researchers. The ones I knew best were Francis Crick and Sydney Brenner. Sydney was the wittiest and most irreverent scientist that I have ever met. He once interviewed me for a job (which I did not get) during which he described his colleagues by comparing them to the crazed figures in Picasso's painting *Guernica*, which hung on the wall of his office. His humour was based on the juxtaposition of the unexpected, and I suspect that was also the source of his immense creativity as a scientist.

These and other code-crackers showed that the four-letter alphabet of DNA is arranged into three-letter 'words' along each strand of the DNA ladder, with most of those short words corresponding to one specific amino acid building block of a protein. The DNA 'word' GCT for example, tells the cell to add an amino acid called alanine to a new protein, whereas TGT would call for an amino acid called cysteine. You can think of a gene as being the sequence of DNA words needed to make a specific protein. For example, a human gene with the name beta-globin contains its essential information in 441 DNA 'letters' (that is, nucleotide bases), which spell out 147 three-letter DNA 'words', which the cell translates into a protein molecule that is 147 amino acids long. In this case, the beta-globin protein helps form the oxygen-carrying pigment called haemoglobin, found in red blood cells, that keeps your body alive and makes your blood look red.

The ability to understand the genetic code solved a key mystery at the heart of biology. It showed how the static instructions stored in the genes could be turned into the active protein molecules that build and operate living cells. Breaking this code paved the way to today's world where biologists can readily describe, interpret and modify gene sequences. At the time, this advance seemed so important that some biologists downed their tools, concluding that the most fundamental problems of cell biology and genetics had now been solved. Even Francis Crick decided to shift his focus from cells and genes to the mysteries of human consciousness.

Today, more than fifty years on, it's clear that things were not quite done and dusted. Nevertheless, biologists had made dramatic progress. Within a century, the gene – which started as an abstract element – had been radically transformed. By the time I finished my PhD in 1973, the gene was no longer only an idea or a part of a chromosome. It was a string of DNA nucleotide bases, encoding a protein with precise functions in the cell.

Biologists soon learned how to find out where particular genes lie on chromosomes, to pluck them out and to move them between chromosomes; even inserting them into the chromosomes of different species. In the late 1970s, for example, the chromosomes of *E. coli* bacteria were re-engineered to contain the human gene that encodes the insulin protein, which regulates blood sugar. These genetically modified, or GM, bacteria produce in affordable abundance a version of the insulin protein that is identical to that made by the human pancreas. They have since helped millions of people around the world manage their diabetes.

During the 1970s the British biochemist Fred Sanger made another crucial innovation when he devised a way to 'read' genetic information. He used an ingenious combination of chemical reactions and physical methods to identify the nature and sequence of all the nucleotide bases that make up a gene (this is called DNA sequencing). The numbers of DNA letters in different genes cover an enormous range, from a couple of hundred bases to many thousands, and the ability to read them and predict the protein they would produce was a great step forward. Fred, who was as extraordinarily modest as he was extraordinarily accomplished, went on to win two Nobel Prizes!

By the end of the twentieth century, entire genomes – that is, the complete set of genes or genetic material present in a cell or organism – could be sequenced, including our own. All three billion DNA letters of the human genome were first sequenced, more or less completely, by 2003. It was a major step forward for biology and for medicine, and progress has not let up since. Whereas sequencing that first genome took a decade and cost more than two billion pounds, today's DNA sequencing machines can do the same in a day or two, for just a few hundred pounds.

The most important thing to come out of the original human genome project was the list of around 22,000 protein-encoding genes, common to all humans, that form the basis of our inheritance. These specify both the features we all share and the inherited characteristics that make us distinct individuals. On its own, that knowledge is not enough to explain what it is to be a human being, but without it our understanding will always be incomplete. It's a bit like having the list of characters in a play – that list is a necessary starting point, but the next, bigger task is to write the play and find the actors that bring those characters to life.

The process of cell division has a vital role in linking together the ideas of 'The Cell' and 'The Gene'. Every time a cell divides, all the genes on all the chromosomes inside that cell must first be copied and then divided equally between the two daughter cells. The copying of the genes and the division of the cell must, therefore, be closely co-ordinated. If they were not, we would end up with cells that would die or malfunction because they lacked the full set of genetic instructions they need. This co-ordination is achieved by the *cell cycle*, the process that orchestrates the birth of every new cell.

DNA is copied early in the cell cycle, during a period of DNA synthesis called S-phase, and the separation of the newly copied chromosomes occurs later, during a process called mitosis. This ensures that the two new cells generated at cell division each have complete genomes. These cell cycle events illustrate an important aspect of life: they are all based on chemical reactions, albeit highly complex reactions. On their own these reactions cannot be considered alive. That only starts to happen when all the hundreds of reactions needed to create a new cell work together to form a whole system that performs a specific purpose. That's what the cell cycle does for the cell: it brings the chemistry of DNA replication to life and in doing so fulfils the *purpose* of reproducing the cell.

I began to recognize the fundamental importance of the cell cycle to understanding life during my early twenties, when I was a graduate student at the University of East Anglia in Norwich, searching for a research project to continue my scientific career. I did not think, however, that the research project I initiated in the 1970s would become my research passion for most of my life.

Like most other processes in the life of cells, the cell cycle is run by genes and the proteins those genes produce. Over the years, my lab's guiding ambition has been to identify the specific genes that run the cell cycle and then find out how they work. To do this we have used fission yeast (a species of yeast which is used to make beer in East Africa), because although it is relatively simple, its cell cycle is fairly similar to the cell cycles seen in many other living organisms, including much larger, multicellular ones like ourselves. We set out to find strains of yeast that contained *mutant* forms of genes involved in the cell cycle.

Geneticists use the word *mutant* in a particular way. A mutated gene is not necessarily aberrant or broken; it simply means a different variant of a gene. The different plant strains that Mendel crossed, such as those with purple or white flowers, differed from each other because of mutations in a gene that is important for determining flower colour. By exactly the same logic, people with differently coloured eyes can be considered as distinct mutant strains of human being. Often it makes no sense to say which of these different variants should be considered 'normal'.

Mutations occur when the DNA sequence of a gene has been altered, rearranged or deleted. This is usually either the result of damage inflicted on the cell – by UV radiation or chemical damage, for example – or due to the occasional mistakes that can occur during the processes of DNA replication and cell division. The cell has sophisticated mechanisms to spot and repair most of these errors, which means that mutations tend to be rather rare. By some estimates, an average of just three small mutations occur each time one of your cells divides: an impressively low error rate of about one per billion DNA letters copied. But once mutations have occurred, they can create different forms of genes that produce altered proteins, which in turn can alter the biology of the cells that inherit them.

Some mutations provide a source of innovation, by changing the way a gene works, occasionally in a useful way, but in many cases mutations stop a gene from carrying out its proper function. Sometimes, the change of just a single DNA letter can have a big effect. For example, when a child inherits two copies of a particular variant of the beta-globin gene, with a change in a single DNA base, their haemoglobin pigment is not fully effective and they develop a blood disorder called sickle cell disease.

To understand how fission yeast cells control their cell cycle, I searched for strains of the yeast that were unable to divide properly. If we could find these mutants, I knew we could then identify the genes required for the cell cycle. My lab colleagues and I started out by looking for fission yeast mutants that could not undergo cell division but could still grow. These cells were quite easy to spot under the microscope because they kept growing without ever dividing and therefore became abnormally enlarged. Over the years, in fact over forty years, the lab has identified more than 500 of these large-celled yeast strains, all of which did indeed turn out to contain mutations which inactivated genes required for specific events in the cell cycle. This means that there are at least 500 genes involved in the cell cycle – that's around

10% of the total set of 5,000 genes found in fission yeast.

This was progress, because these genes were clearly needed for a yeast cell to complete the cell cycle. However, they did not necessarily *control* the cell cycle. If you think about the way a car works, there are many components that will stop a car when they break: the wheels, the axles, the chassis and the engine, for instance. These are all important, to be sure, but none of them are used by the driver to control the speed of the car's travel. Returning to the cell cycle, what we really wanted to find were the accelerator, gearbox and brakes; that is, the genes that control how *quickly* cells progress through the cell cycle.

In the event, I stumbled across the first of these cell cycle control genes entirely by accident. I remember vividly the moment in 1974 when I was using a microscope to search laboriously for yet more colonies of abnormally enlarged mutant yeast cells – this was quite a chore because only about 1 in every 10,000 colonies I looked at was of any real interest. It took a whole morning or afternoon to find each of these mutants, and some days I didn't find any at all. Then I noticed a colony that contained cells which were unusually small. At first I thought they might be bacteria that had contaminated my Petri dish, a fairly common frustration. Looking more carefully, I realized that they could actually represent something more interesting. Perhaps they were yeast mutants that raced through the cell cycle before they had time to grow, and therefore divided at a smaller size?

This line of thinking turned out to be correct; the mutant cells were indeed altered in a gene that controlled how quickly a cell could undergo mitosis and division, and so complete its cell cycle. This was exactly the kind of gene I was hoping to find. These cells really were a bit like cars with a defective accelerator that makes the car, or in this case the cell cycle, go faster. I called these diminutive strains 'wee' mutants, since they were isolated in Edinburgh, and 'wee' is the Scottish word for small. I must confess that the wit wears thin after half a century!

The gene altered in that first wee mutant turned

out to work with another even more important gene, one at the very heart of cell cycle control. As things happened, another good dose of happenstance led me to find that second elusive control gene too. I had been working for many months isolating different strains of small-celled wee mutants and had painstakingly gathered nearly 50 of them. This was an even bigger slog than looking for the abnormally large-celled mutants: it took nearly a week to find each one. This challenge was compounded by the fact that most of the strains I laboured to identify were of limited interest because they all contained subtly different mutations of the same gene, which I had by then called *wee1*.

Then, one wet Friday afternoon, I spotted another *wee* mutant. This time my Petri dish was definitely contaminated: the dish, and the abnormally small yeast cells that had caught my eye, were covered by the long tendrils of an invading fungus. I was tired and knew that getting rid of such a contaminating fungus was a long and tedious task. In any case, I assumed this new strain would most likely contain yet another mutant form of the same gene, *wee1*. I threw the whole Petri dish into the rubbish bin and went home for my tea.

Later that evening I felt guilty about what I had done. What if *this* mutant was different from the other 50 wee mutants? By then it was a particularly dark and wet Edinburgh night, but I got back on my bicycle and rode back up the hill to the lab. Over the next few weeks I managed to isolate the new wee mutant away from the invading fungus. And then – to my sublime pleasure – it turned out that this wasn't yet another variant of the *wee1* gene. It was a completely new gene and, ultimately, the key that unlocked how the cell cycle was controlled.

I called my new gene cell division cycle 2, or *cdc2* for short. Looking back, I sometimes wish I'd given this central part of the cell cycle puzzle a more elegant, or at least a more memorable name! Not least since you're going to hear rather more about *cdc2* later in this book.

With the benefit of hindsight, all of this was really quite simple, both to do and to think about. Luck was very important too: both the acciden-tal finding of the first wee mutant, which I was not even searching for, and the quirk of fate that meant the 'failed' experiment I retrieved from the rubbish bin was the one that eventually led me to the central player in cell cycle control. Simple experiments and thinking can be surprisingly illuminating in science, especially when combined with a good measure of hard graft, hopefulness, and, of course, the occasional lucky break.

I did most of these experiments when I was a junior scientist, with a young family at home, working in the lab of Professor Murdoch Mitchison in Edinburgh. He provided the space and equipment I needed to do my experiments, as well as an endless supply of advice and comment on what I was doing. Despite all his input, he would not let me include him as an author on any of my papers because he did not think he had contributed enough. It was not true, of course. It is generosity like that which has been my principal experience of doing science, but it gets less attention than it should. Murdoch was an interesting man. Generous, as I have said, somewhat shy, and utterly consumed by his research. He cared little about whether others were interested in what he was doing; he marched to the beat of his own drum. If Murdoch was still around, he might not have approved of my singling him out like this here, but I want to give him full credit for showing me why the best research is both intensely individual and utterly communal.

Life cannot exist without genes: each new generation of cells and organisms must inherit the genetic instructions they need to grow, function and reproduce. This means that for living things to persist in the long term, genes must be able to replicate themselves very precisely and carefully. Only that way can the DNA sequences be kept constant through multiple cell divisions, so genes can withstand the 'test of time'. Cells achieve this with impressive exactitude. We see the result of this all around us. The DNA sequence of the huge majority of the 22,000 genes that control your cells is almost completely identical to those of all other people on this planet today. They are also largely indistinguishable from those of our ancestors who hunted, gathered and swapped stories around campfires in the depths of pre-history, tens of thousands of years ago. Altogether, the mutations that differentiate your inborn characteristics from mine, and both of us from our prehistoric ancestors, add up to a tiny fraction – less than one per cent – of your total complement of DNA code. This is one of the big discoveries of twenty-first century genetics: our genomes, each three billion DNA 'letters' long, are very similar, across genders, ethnicities, religions and social classes. This is an important equalizing fact that societies across the world should appreciate.

We cannot disregard those scattered variations that we all carry in our genes, however. Although in the small minority overall, they can have a big effect on our individual biology and life history. Some of these variants are shared between me and my daughters and grandchildren, and they explain some aspects of our resemblance as a family. Other gene variants are unique to each of us, and are part of what makes us into distinct individuals, by influencing our physical appearance, our health and our ways of thinking, in either subtle or not so subtle ways.

Genetics is central to all our lives, shaping our sense of identity and outlook on the world. Late in my life, I discovered something rather surprising about my own genetics. I grew up in a working-class family; my father worked in a factory and my mother was a cleaner. My brothers and sister all left school when they were fifteen, so I was the only one who stayed on at school and, later, went on to university. I had a happy and well-supported, if somewhat old-fashioned, childhood. My parents were rather older than those of my friends, and I used to quip that it was like being brought up by my grandparents.

Many years later I applied for a 'Green Card' so I could take up residence in the USA and start my new job as President of the Rockefeller University, in New York. To my surprise, my application was rejected. The US Department of Homeland Security said it was because the version of the birth certificate I had used all my life did not list the names of my parents. Irritated, I wrote off for the full version of my birth certificate. The shock came when I opened the envelope containing that new certificate. What it showed was that my parents were not my parents – they really *were* my grandparents. My mother was actually my sister. It turned out that she had got pregnant at seventeen and, since illegitimacy was considered rather shameful at that time, she had been sent to her aunt's home in Norwich, which is where I was born. When we returned to London, my grandmother, wanting to protect her daughter, pretended to be my mother, and brought me up. The great irony on discovering all this was that although I am a geneticist, I did not know my own genetics! In fact, because everyone who might have known has since died, I still don't know who my father is: there is just a dash on my birth certificate where his name should be.

All individuals are born with a relatively small number of novel genetic variants that tend to arise at random and are not shared with either of their biological parents. As well as contributing to what makes individual organisms unique, these heritable differences also explain why living species are not static and unchanging over long periods of time. Life is constantly experimenting, innovating and adapting as it changes the world and the world changes around it. For this to be possible, genes must balance the need to preserve information by staying constant, with the simultaneous ability to change, sometimes substantially so. The next idea shows us how this can come about and, as a result, how life became so bewilderingly diverse.

That idea is evolution by natural selection.

3. EVOLUTION BY
   NATURAL SELECTION

Chance and Necessity

The world is teeming with an extraordinary diversity of life forms. The yellow butterfly that started this book was a brimstone, an early harbinger of spring. With its delicate yellow wings, it is a beautiful example of the amazingly diverse group of animals that we call insects.

I like insects, particularly beetles, which were a hobby of mine when I was a teenager. There is an astonishing variety of beetles – some scientists think there are over one million distinct species of them throughout the world. Growing up in England, I marvelled at armour-plated ground beetles scurrying around under stones, beetles that glowed at night, red and black ladybirds eating aphids in the garden, powerful water beetles swimming in ponds and weevils in the flour packet. Beetles present us with a cacophony of diversity; they are a microcosm of the diversity of all life.

Life in all its different forms can at times seem overwhelming: we share our world with countless animals, birds, fish, insects, plants, fungi and an even longer roster of different microbes, each appearing to be well adapted to their own particular lifestyle and environment. No wonder that for millennia most people thought that all this diversity must have resulted from the efforts of a divine Creator.

Creation myths abound in most cultures. The Judaeo-Christian myth of Genesis, if read literally, claims life was created within just a few days. The pervasive idea, that individual species had each been fashioned by a Creator, led the twentieth century geneticist J. B. S. Haldane to look at the huge diversity of beetles and quip that whoever God is, 'He has an inordinate fondness for beetles.'

During the eighteenth and nineteenth

centuries, thinkers began to compare the intricate mechanisms of living things with those of the complex machines being designed and constructed during the Industrial Revolution. These comparisons often reinforced religious beliefs: how could such intricacy have come to be without the input of a supremely intelligent designer?

One colourful example of this kind of reasoning came from the Reverend William Paley in 1802. He asked you to imagine that you were out walking and found a watch on the path. If you opened the watch and examined its complex mechanism, clearly designed for the purpose of tracking time, it would, he argued, convince you that the watch was made by an intelligent Creator. According to Paley, the same logic must apply to intricate living mechanisms.

We now know that complex life forms endowed with a sense of purpose can be generated without a designer of any kind, and that is due to natural selection.

Natural selection is the intensely creative process that has produced us – and the extraordinary diversity of living forms that surrounds us – from the millions of different microbe species to the fearsome jaws of the stag beetle, the 30-metre tentacles of the lion's mane jellyfish, the fluid-filled traps of the carnivorous pitcher plant and the opposable thumbs of the great apes, including ourselves. Without ever deviating from the laws of science or invoking supernatural phenomena, evolution by natural selection has generated populations of increasingly complex and diverse creatures. Over aeons of time, different species have risen to prominence, their forms changing beyond recognition, as they have explored new possibilities and interacted with different environments and other living creatures. All species – including our own – are in a state of perpetual change, eventually becoming extinct or developing into new species.

For me this story of life is just as full of wonder as any of the creationist myths. Whereas most of the religious stories present us with creative acts that are familiar, even somewhat mundane, and durations of time that we can readily understand, evolution by natural selection pushes us to imagine something much more at the edge of our comfort zone, but also more magnificent. It is a wholly undirected and incremental process, but when it is embedded in the inconceivably vast duration of time, what scientists sometimes call 'deep time', it becomes the most supremely creative force of all.

The towering figure in evolution is Charles Darwin, the nineteenth-century naturalist who travelled the world in the tiny Royal Naval ship *HMS Beagle*, collecting specimens of plants, animals and fossils. Hungrily, Darwin gathered observations that supported the idea of evolution and came up with a beautiful mechanism – natural selection – that explained how it worked. He shared all of this in his 1859 book *On the Origin of Species*. Of all the great ideas of biology, this is probably the best known, if not always the best understood.

Darwin was not the first to suggest that life evolved over time. As he notes in *On the Origin of Species*, Aristotle had argued that body parts of animals might appear or disappear over long periods of time. The late eighteenth-century French scientist Jean-Baptiste Lamarck took this further, arguing that different species were linked together in chains of relatedness. He proposed that species change gradually through the process of adaptation, with their form responding to shifts in the environment and changes in their habits. Famously, he argued that giraffes developed their long necks because, with each generation, they stretched upwards to reach leaves higher up on trees, and somehow, the results of that exertion were passed on to their offspring, who would have slightly longer necks. Lamarck's ideas are sometimes belittled today because he did not get the details of the process of evolution right, but he deserves great credit for providing one of the first comprehensive accounts of the phenomenon of evolution, if not its cause.

Lamarck was certainly not alone in speculating about evolution. Even in Charles's own family, his colourful grandfather, Erasmus Darwin, was another early and enthusiastic supporter of evolution. He had a motto inscribed on his coach which read 'E conchis omnia', that is 'everything from shells', advertising his belief that all life developed from much simpler ancestors, such as the apparently formless blob of a mollusc inside its shell. However, he had to remove it after the Dean of Lichfield Cathedral accused him of having 'renounced his Creator'. Erasmus obliged, since he was also a successful doctor and understood that, had he not done so, he would have been in danger of losing his more respectable, and therefore wealthier, patients. He was also considered at the time to be a distinguished poet, expounding his views on evolution in verses from his poem *The Temple of Nature*:

'First forms minute, unseen by spheric glass
Move on the mud, or pierce the watery mass;
These, as successive generations bloom,

New Powers acquire and larger limbs assume;
Whence countless groups of vegetation spring
And breathing realms of fin, and feet, and wing.'

His reputation as a poet may not have survived, perhaps understandably, but his reputation as a scientist has. However, his lines anticipated aspects of the ideas elaborated by his better-known grandson.

Charles Darwin was more scientific and systematic in his approach to evolution, and his means of communication were more conventional, confining himself to prose rather than verse. He amassed huge amounts of observational data from the fossil record and his studies of plants and animals, both at home and abroad. He organized it all to provide strong evidence for the view, shared by Lamarck, his grandfather and others, that living organisms *do* evolve. But Darwin did more than that when he proposed natural selection as a *mechanism* for evolution. He joined up all the dots and showed the world how evolution could actually work.

The idea of natural selection is based on the fact that populations of living organisms exhibit variations, and when these variants are caused by genetic changes, they will be inherited from generation to generation. Some of these variants will affect characteristics that make certain individuals more successful in producing offspring.

This enhanced reproductive success means that the offspring possessing these variants will make up a greater proportion of the population in the next generation. In the case of the giraffe's long neck, we can infer that the random appearance and accumulation of variants that subtly altered the skeleton and muscles of the neck allowed some of the giraffe's ancestors to reach slightly higher branches, eat more leaves and gain more nutrition. Eventually, those that could do so proved more resilient and more capable of producing young giraffes, so the herds of giraffes roaming the savannahs of Africa gradually became dominated by individuals with longer necks. This process is known as *natural* selection since constraints imposed by all manner of *natural* factors, such as competition for food or mates or the presence of diseases and parasites, ensure that some individuals fare better and therefore reproduce more than others.

The same mechanism was put forward independently by the naturalist and collector Alfred Wallace. It's less widely known that both of them followed speculations about natural selection made earlier in the century, in particular by the Scottish agriculturalist and landowner Patrick Matthew in his 1831 book on naval timber. Nevertheless, Darwin was the first to present the whole idea in a convincing, comprehensive and enduringly compelling way.

Humans have actually been hijacking the same process for thousands of years, using it to breed organisms possessing particular characteristics. This is called *artificial* selection, and Darwin actually developed his ideas about natural selection by observing the way pigeon fanciers selected particular individuals to breed to produce a wide range of pigeon varieties. Artificial selection can produce dramatic results. It is how we transformed wild grey wolves into man's best friend, creating dog breeds that range from the tiny Chihuahua to the towering Great Dane. It's also how the wild mustard plant gave rise to broccoli, cabbage, cauliflower, kale and kohlrabi. These transformations have taken place over a relatively modest number of generations, giving a glimpse of the great power of the evolutionary process when it is allowed to run its course naturally over millions of years.

Natural selection leads to survival of the fittest – which, incidentally, is not a term Darwin himself used – and to the elimination of individuals that cannot compete. As a consequence of this process, specific genetic changes accumulate in populations, resulting ultimately in enduring changes to the form and function of living species. It can explain how some beetles developed red-spotted wing cases, whilst others learned to swim, roll balls of dung, or glow in the dark.

Natural selection is a profound idea, which has significance beyond biology. It has both explanatory power and practical utility in several other disciplines, not least economics and computer science. Today, for example, some aspects of software and some engineered components of machines, such as aircraft, are optimized by algorithms that mimic natural selection. These products are evolved, rather than designed in the traditional sense.

For evolution by natural selection to take place, living organisms must have three crucial characteristics.

First, they must be able to reproduce.

Second, they must have a hereditary system, whereby information defining the characteristics of the organism is copied and inherited during their reproduction.

Third, the hereditary system must exhibit variability, and this variability must be inherited during the reproductive process. It is this variability that natural selection operates upon. It transforms a slow and randomly generated source of variability into the apparently boundless and constantly changing range of life forms that flourish around us.

Additionally, for this to work efficiently, living organisms must die. That way, the next generation, potentially containing genetic variants that give them a competitive edge, can replace them.

The three necessary characteristics emerge directly from the ideas of the cell and the gene. All cells reproduce during the cell cycle and all cells have a hereditary system made up of genes, which are copied and inherited on the chromosomes during mitosis and cell division. Variation is introduced by the appearance of chance mutations that change DNA sequences – like the one that led me to discover the *cdc2* gene – which result from either rare mistakes during the copying of the double helix, or environmental damage to the DNA. Cells repair these mutations, but they are not completely successful. If they were, all individuals of a species would be identical and evolution would stop. This means the error rate itself is subject to natural selection. If that error rate is too high the information stored by the genome will degenerate and become meaningless, and if errors are too rare, the possibility for evolutionary change is reduced. Over the long term, the most successful species will be those that can maintain the right balance between constancy and change.

In complex eukaryote organisms, further variability is introduced during the process of sexual reproduction, when parts of chromosomes are reshuffled during the cell divisions that produce sex cells (also known as germ cells: sperm cells and egg cells in animals, and pollen and ovules in flowering plants), which are made by the process called *meiosis*. That is the main reason siblings are genetically different from each other: if their parents' genes are like a deck of cards, they are each dealt a different genetic 'hand'.

Many other organisms introduce variation by exchanging DNA sequences directly, between different individuals. This is common in less complex organisms like bacteria, which can swap genes between one another, but also with more complex organisms. This process is called horizontal gene transfer. It is one of the reasons the genes that make certain bacteria resistant to antibiotics can spread rapidly through whole populations of bacteria, and even from one unrelated species to another. Horizontal gene transfer also makes it harder to trace some lineages back through evolutionary time, since it means that the inheritance of genes can flow from one branch of the tree of life into another.

Whatever the source of genetic variability, to fuel evolutionary change it must persist during subsequent reproduction and generate populations of organisms that vary across every possible dimension, whether that's subtle differences in disease resistance, attractiveness to mates, food tolerance, or any number of other characteristics. Natural selection can then act to sift the helpful variants from the harmful.

One profound consequence of evolution by natural selection is that all life is connected by descent. This means that as the tree of life is traced backwards, the branches increasingly converge into bigger branches and eventually into a single trunk. The conclusion therefore, is that we humans are related to every other life form on the planet. To some, like the apes, we are closely related, because we are on adjacent twigs at the edge of the tree, and to others, like my yeast, the relationship is much more distant, because we only become 'joined' much further back in time, closer to the main trunk of the tree of life.

Our fundamental connectedness to other life was brought home to me when I went trekking through the humid and verdant Ugandan rainforest, in search of mountain gorillas. Following my guide, we suddenly came upon a family group. I found myself sitting opposite a magnificent silverback, who was squatting underneath a tree, just two or three metres away from me. I broke out into a sweat, and it was not just because it was hot and humid. As a geneticist, I knew that he and I shared about 96% of our genes, but that bald number can only tell part of the story. As his intelligent, deep brown eyes locked my gaze, I saw many aspects of my humanity reflected back at me. Those apes were closely attuned to each other and also to us humans. Much of their behaviour was inescapably familiar; their empathy and curiosity obvious. The silverback and I contemplated each other for several minutes. It was like a conversation. Then he put out a hand, bent double a five-centimetre diameter sapling (was he trying to tell me something?) and slowly climbed the tree, all the time holding me in his gaze with those penetrating eyes. This dramatic and moving encounter emphasized for me quite how closely we are related to these magnificent creatures. That connectedness extends beyond the gorilla to other apes, to mammals and other animals, and even, via more ancient forks in life's shared family tree, to plants and microbes. This, to me, is one of the best arguments for why humanity should care about the entire biosphere; all the different life forms that share our planet are our relatives.

I became aware of our deep relatedness to other living things in an even more unexpected way when I decided to ask whether fission yeast and human cells controlled their cell cycles in the same way. I asked this question during the 1980s, when I found myself working in a cancer research institute in London. Since cancer is caused by aberrant cell division of human cells, most of my colleagues, working in other labs, very understandably, were much more interested to know what controlled the cell cycle in humans rather than in yeast. By that time I knew what controlled cell division in yeast: it was a cell cycle control mechanism with *cdc2*, that crucial gene with the uninspiring name, at its very centre.

I wondered if it could possibly be the case that human cell division was also controlled by a human version of the same gene, *cdc2*? This seemed very unlikely, given that yeasts and humans are so very different and last had an ancestor in common 1.2 to 1.5 billion years ago (that is, 1,200 to 1,500 million years ago). To put that huge expanse of time in perspective, dinosaurs became extinct a 'mere' 65 million years ago, and the first simple animals appeared around 500–600 million years ago. If I'm completely honest, it was more than slightly preposterous to believe that such distant relatives could have cells whose reproduction was controlled in the same way. Nevertheless, we had to find out.

The way Melanie Lee in my lab tackled this question was to try and find a human gene that functioned the same way as did *cdc2* in fission yeast. To do this, she took fission yeast cells that were defective in *cdc2* and so could not divide, and 'sprinkled' on them a gene 'library' that was made up of many thousands of pieces of human DNA. Each piece of DNA contained a single human gene. Melanie used conditions that ensured that the mutant yeast cell would usually only take up one or two genes. *If* it so happened that one of these genes was the human equivalent of the *cdc2* gene and *if* it functioned in the same way in both human and yeast cells, and *if* the human *cdc2* gene could get into the yeast cells, then the *cdc2* mutant cells might just regain the ability to divide. *If* all that went right, they should form colonies that Melanie could see on a Petri dish. You may have noticed there were several 'ifs' in this plan. Did we think the experiment would work? Probably not, but it was worth a shot.

And, amazingly, it did work! Colonies grew on the Petri dish and we were able to isolate the stretch of human DNA that had successfully stood in for the *cdc2* gene that is so vital for yeast cell division. We sequenced this unknown gene and saw that the sequence of the protein it made was very similar to the yeast Cdc2 protein. It was obvious that we were looking at two highly related versions of the same gene. So similar were they that the *human* gene could control the *yeast* cell cycle.

This unexpected result led us to a far-reaching conclusion. Given that fission yeast and humans are so distantly related in evolution, it was very likely that cells in every animal, fungus and plant on the planet controlled their cell cycle in the same way. They almost certainly all depended on the action of a gene that was very similar to yeast's *cdc2* gene. And, what is more, even as these different organisms gradually evolved over aeons of evolutionary time to take on countless different forms and lifestyles, the core controls of this most fundamental process had barely changed. *Cdc2* is an innovation that has endured for more than one billion years.

All of this reinforced my conviction that understanding how human cells control their division, which is crucial for understanding how our bodies change as we grow, develop, suffer diseases and degenerate across our lifespans, can be studied profitably in a wide range of living organisms, including the simple yeast.

Natural selection not only takes place during evolution; it is also taking place at the level of the cells inside our bodies. Cancer starts when genes important for controlling the growth and division of cells are damaged or rearranged, leading to cells that divide in uncontrolled ways. Just like evolution within a population of organisms, these pre-cancerous or cancerous cells can, if they evade the body's defences, gradually overtake the population of unaltered cells that make up the tissue. As the population of damaged cells grows, there is a greater chance of further genetic changes taking place within these cells, leading to an accumulation of genetic damage and the generation of increasingly aggressive cancerous cells.

This system has the three characteristics necessary for evolution by natural selection: reproduction, a hereditary system, and the ability of the hereditary system to exhibit variability. It is paradoxical that the very circumstances that allowed human life to evolve in the first place are also responsible for one of the most deadly human diseases. More practically, it also means that population and evolutionary biologists should be able to contribute significantly to our understanding of cancer.

Evolution by natural selection can bring about great complexity and the apparent purposefulness of living things. It does this without any controlling intellect, defined end goal, or ultimate driving force. It sidesteps completely the arguments invoking a divine Creator, as made by Paley with his imagined pocket watch, and many others before and since. And it leaves me, for one, in a state of constant wonder.

Learning about evolution also had a rather dramatic impact on the course of my life. My grandmother was a Baptist, so we used to go to the local Baptist church every Sunday. I knew the Bible well (still do), and even had thoughts of becoming a minister, perhaps even a missionary! Then, around the time I saw that brimstone butterfly in my garden, I was taught about evolution by natural selection at school. The scientific explanation for life's abundant diversity was clearly in direct conflict with the biblical account. Trying to make sense of this discrepancy, I went to talk to my Baptist minister. I suggested to him that when God was speaking about the Genesis account of creation, he was explaining what happened in terms that would make sense to an uneducated, pastoral population two or three thousand years ago. I said that we should perhaps treat it more like a myth, but that, in reality, God had devised an even more wonderful mechanism for creation, by inventing evolution by natural selection. Unfortunately my minister did not see it like that at all. He told me I had to believe the literal truth of Genesis and said that he would pray for me.

Thus began my gradual descent from religious belief to atheism, or to be more precise, sceptical agnosticism. I saw that different religions can have very different beliefs, and that those different creeds could be inconsistent with each other. Science gave me a route to a more rational understanding of the world. It gave me greater certainty too, stability even, and a better way to pursue truth; the ultimate objective of science.

Evolution by natural selection describes how different life forms can come about and attain purpose. It is driven by chance and steered by the necessity of generating increasingly effective life forms. However, it does not give much insight into how living organisms actually work. For that we have to turn to the next two ideas, the first of which is *life as chemistry*.

4. LIFE AS CHEMISTRY

Order from Chaos

Most people would probably look at the world around them and divide it into two main types of thing: those that are alive and those that are clearly not. Living organisms stand out because they are things of action; they behave with purpose, reacting to their surroundings and reproducing themselves. None of these characteristics apply to things that are not living, like a pebble, a mountain, or a sandy beach, for example. Indeed, if we were to step back in time a couple of hundred years, before the development of the ideas described in this book, we might well have concluded that earthly life is directed by mysterious forces that are unique to living things.

This way of thinking is called 'vitalism', and its origins trace back to the classical thinkers Aristotle and Galen, and probably even further. Even for the most rational and scientific among us, it is hard to entirely abandon such thinking. If you've ever seen someone die, you will know it can seem very much as if some inexplicable spark of life has been abruptly extinguished.

Vitalist explanations are appealing since they seem to provide a comforting solution to what our minds struggle to grasp. But we can in fact now be certain that we do not need to invoke any form of magic. Most aspects of life can be understood rather well in terms of physics and chemistry, albeit an extraordinary form of chemistry that is highly ordered and organized, and of a sophistication that cannot be matched by any inanimate process. For me, this explanation is more awe-inspiring than any kind of belief that life is directed by mysterious forces that lie beyond the reach of scientific scrutiny.

This idea that life *is* chemistry rather surprisingly had its origins in studies of fermentation, the process by which the simple microbe yeast makes alcohol during the production of beer and wine. This has been a long-standing interest of humanity.

In fact, my own life has been influenced quite a lot by fermentation, and not just because I am rather fond of beer myself; sitting alone contemplating the world in an empty pub early in the evening is a real pleasure. When I left school at seventeen years of age, I knew I wanted to study biology, but I couldn't get a place at university. At that time a basic foreign language qualification, achieved via an exam known as an O-level, was a compulsory entry requirement for all undergraduate degree programmes, but I managed to fail my French exam six times, probably a world record in O-level failures! So I didn't go to university, but went to work instead as a technician in a microbiology laboratory linked to a brewery.

Part of my job each day was to make all the nutrient-containing concoctions that the scientists needed to grow their microbes. I soon realized that they nearly always placed the same daily order, so I could make it all up in a big batch on Monday, which would then last all week. I went to see my boss Vic Knivett (who, incidentally, was a Georgian dancer in his spare time, a fact I discovered when I found him one evening performing an energetic Cossack-like kicking dance on top of one of the laboratory benches!). Generously, he suggested I carry out a research project on *Salmonella* infections of hens' eggs. I was an eighteen-year-old in heaven, doing experiments every day, pretending to be a real scientist.

At some point during that year in the brewery, a sympathetic professor at Birmingham University called me for an interview, and eventually persuaded the university to overlook my weaknesses in foreign languages so I could start a biology degree in 1967. Ironically, considering my early struggles with the language, thirty-five years later I was granted an award called the Légion d'honneur by the President of France for my research on yeast. I even had to give my acceptance speech in French! However, despite studying yeast for most of my life, I have never made a drop of either wine or beer myself.

The scientific study of fermentation began with the eighteenth-century French nobleman and scientist Antoine Lavoisier, one of the founders of modern chemistry. Unfortunately for him, and unfortunately for science as a whole, his part-time activities as a tax collector meant he lost his head in May 1794, during the French Revolution. The judge of the kangaroo political court that sentenced him declared that the 'Republic has no need for savants and chemists'. We scientists obviously have to treat politicians with caution! There is an unfortunate tendency for politicians, especially those of a populist bent, to ignore 'experts', particularly when that expertise counters their poorly substantiated opinions.

Before his untimely encounter with the guillotine, Lavoisier had become fascinated by the process of fermentation. He concluded that 'fermentation was a *chemical reaction* in which the sugar of the starting grape juice was converted into the ethanol of the finished wine'. Nobody had thought about it in quite that way before. Then Lavoisier went further and proposed that there was something called a 'ferment', which seemed to come from the grapes themselves, which played a key role in the chemical reaction. He couldn't say what this 'ferment' was, however.

Things became clearer about half a century later, when the makers of industrial alcohol asked the French biologist and chemist Louis Pasteur to help them solve a mystery that kept destroying their products. They wanted to know why their fermentations of sugar beet pulp sometimes went wrong, producing a sour and unpleasant acid instead of ethanol. Pasteur embarked on this mystery as a detective might. Using a microscope, he obtained the crucial clue. Sediments in the fermentation vats that produced alcohol contained yeast cells. The yeast were clearly alive because some of them had buds, showing they were actively multiplying. When he looked into the soured vats he couldn't see any yeast cells at all. From these simple observations, Pasteur proposed that the microbial life form yeast was the elusive ferment: the key agent responsible for making ethanol. Some other microbe, probably a smaller bacterium, generated the acid that ruined the failed batches.

The point here was that the growth of living cells was directly responsible for a specific chemical reaction. In this case, the yeast cells were converting glucose into ethanol. The most important thing Pasteur did was to step from the specific to the general, to reach an important new conclusion. He argued that chemical reactions were not just an interesting feature of cellular life – they were one of life's defining features. Pasteur summarized this brilliantly when he said that 'chemical reactions are an *expression of the life of the cell*'.

We now know that within the cells of all living organisms many hundreds, even thousands, of chemical reactions are being carried out simultaneously. These reactions build up the molecules of life, which form the components and structures of cells. They also break molecules down, to recycle cellular components and release energy. Together, the vast array of chemical reactions occurring in living organisms is called *metabolism*. It is the basis of everything living things do: maintenance, growth, organization and reproduction, and the source of all the energy needed to fuel these processes. Metabolism is the chemistry of life.

But how are all the many and varied chemical reactions that make up metabolism brought about? What type of substance was it in Pasteur's yeast that was carrying out the chemical reactions of fermentation? Another French chemist, Marcelin Berthelot, delved deeper into this mystery and made the next advance. He pulverized yeast cells, and from the cellular remnants extracted a chemical substance which behaved in an intriguing way. It triggered a specific chemical reaction – the conversion of table sugar, sucrose, into its two smaller component sugars: glucose and fructose – but it was not itself consumed by the reaction. It was an inanimate substance but integral to a living process and, notably, it continued to work when it was removed from the cell. He called this new substance invertase.

Invertase is an enzyme. Enzymes are catalysts: that means they facilitate and speed up chemical reactions, often dramatically. They are acutely important for life. Without them many of the chemical processes most vital for life would simply not happen, especially at the relatively low temperatures and mild conditions found within most cells. The discovery of enzymes laid the foundations for today's consensus view – shared by all biologists – that most phenomena of life can be best understood in terms of chemical reactions that are catalyzed by enzymes. To understand how enzymes achieve this we need to understand what they are and how they are made.

Most enzymes are made from proteins, which are built by the cell as long, chain-like molecules called polymers. Polymer structures are fundamentally important for every aspect of life's chemistry. As well as most enzymes and all other proteins, all the lipid molecules that make up cell membranes, all the fats and carbohydrates that store energy, and the nucleic acids responsible for heredity, deoxyribonucleic acid (DNA) and the closely related ribonucleic acid (RNA), are all polymers.

These polymers are built principally from the atoms of just five chemical elements: carbon, hydrogen, oxygen, nitrogen and phosphorus. Of these five, carbon plays a particularly central role, largely because it is more versatile than the other elements. Whereas hydrogen atoms, for example, make only one connection – that is, a chemical bond – with other atoms, each carbon atom can bond to four other atoms. This is the key to carbon's polymer-making abilities: two of carbon's four potential bonds can be linked to two other atoms, often other carbon atoms, creating a chain of linked atoms – the core of each polymer. That leaves each carbon with two further bonds available to link with other atoms. These extra bonds can then be used to add other molecules to the sides of the main polymer chain.

Many of the polymers found in cells are very large molecules, so large, in fact, that they are given a special name: macromolecules. To get a sense of quite how big these molecules can be, remember that the DNA macromolecules at the core of each of your chromosomes can be several centimetres long. That means they incorporate millions of carbon atoms into an incredibly long, but incredibly slender, molecular thread.

Protein polymers are not so long, generally being based on a few hundred to several thousand linked carbon atoms. However, they are far more chemically variable than DNA, which is the main reason they can work as enzymes and therefore play a dominant role in metabolism. Each protein is a carbon-based polymer built by joining smaller amino acid molecules together, one at a time, into a long chain. Invertase, for example, is a protein molecule made by linking together 512 amino acids in a specific, ordered sequence.

Life uses twenty different amino acids. Each of these amino acids have side molecules, branching off from the main polymer chain, that grant them distinct chemical properties. For example, some amino acids have positive or negative charges, others are either attracted to or repelled by water, and some are capable of easily forming bonds with other molecules. By stringing together different combinations of amino acids, each with different side molecules, cells can create a huge range of different protein polymer molecules.

Then, once these linear protein polymer

chains are assembled, they fold, twist and combine together to create complex three-dimensional structures. It's a bit like the way a length of sticky tape can wrap itself up into a tangled ball, although the way proteins fold up is a much more repeatable process that generates a very precise structure. In a cell, the same string of amino acids will always try to form the same specific shape. This leap from the one-dimensional to three-dimensional is crucial, since it means each protein has a distinctive physical shape and a unique set of chemical properties. As a result cells can build enzymes in such a way that they fit together very precisely with the chemical substances they work on – parts of the invertase and sucrose molecules are a perfect fit, for example. This in turn allows enzymes to provide the precise chemical conditions needed to bring about specific chemical reactions.

Enzymes execute almost all the chemical reactions that form the basis of cellular metabolism. But as well as building other molecules up and breaking them down, they play many other roles too. They act as quality controllers, they ferry components and messages between different regions of the cell, and they transport other molecules in and out of the cell. Others still are on the lookout for invaders, activating the proteins that defend cells and therefore our bodies from disease. And enzymes are not the only type of protein. Nearly every part of our bodies – from the hairs on our heads to the acid in our stomachs and the lenses in our eyes – are either made from protein, or are constructed by proteins. All these different proteins have been honed by millennia of evolution to fulfil specific functions within the cell. Even a relatively simple cell contains a huge number of protein molecules. In total, there are over 40 million of them in a tiny yeast cell; that's twice as many proteins in a minuscule cell as there are people in a gigantic city like Beijing!

The outcome of all of this protein diversity is a maelstrom of chemical reactions being carried out in every cell at all times. If you could imagine looking inside a living cell with eyes that could perceive the molecular world, your senses would be assaulted by a boiling tumult of chemical activities.

Some of the molecules involved are electrically charged, making them either sticky or repellent, while others are passively neutral. Some are acids or bleach-like alkalis. All of these different substances are constantly interacting, either through random collisions or stage-managed meetings. Sometimes molecules come together transiently to react chemically, through a quick exchange of electrons or protons. At other times molecules remain chemically connected through the formation of tight and enduring bonds. Altogether, the cell contains many thousands of different chemical reactions that are constantly working away to sustain life. The number of chemical reactions used in even the largest industrial chemical plant pales in comparison. A plastics factory, for example, might be based on a few tens of chemical reactions.

All this frantic and fast activity occupies the opposite end of the time spectrum from the deep time that was needed for these systems to evolve. But the dizzying timescale of the cellular world is just as challenging for our brains to comprehend as evolutionary time. Some of the cell's enzymes that control these reactions work at an astonishingly fast rate, rattling through thousands, even millions, of chemical reactions every second. These enzymes are not only extremely rapid, but can also be extremely precise. They can manipulate individual atoms with a level of accuracy and reliability that chemical engineers can only dream of. But then evolution has been working to refine these processes for billions of years – rather longer than us humans!

Making all this work together is an extraordinary achievement. Although the vast array of chemical reactions that occur simultaneously in cells may appear chaotic, it is in fact very highly ordered. For them to function properly, each of the different reactions requires its own particular chemical conditions. Some require more acid or alkaline surroundings; others demand particular chemical ions like calcium, magnesium, iron or potassium; others need or are slowed down by the presence of water. Yet somehow all these different chemistries must be carried out both simultaneously and very close together in the narrow

confines of the cell. This is only possible because
the various enzymes do not each require different
extreme temperatures, pressures or acid or alkaline conditions, such as those found in industrial
chemical facilities. If they did, they would not all
be able to co-exist in such close proximity. Many of
these metabolic reactions still need to be kept separate from one another, however. They must not
interrupt each other, and all their specific chemical
requirements must be met. Key to answering this
challenge is *compartmentation.*

Compartmentation is a way to get complex
systems of all kinds to work. Take cities. They only
work efficiently if they are organized into different
compartments with particular functions: railway
stations, schools, hospitals, factories, police stations, power stations, sewage disposal plants, and
so on. All of these and much more are required for
the city to work as a whole, but everything would
break down if they were all completely mixed up
together. They have to be separate to work effectively, but they also need to be relatively close
together and connected. It is just the same for

cells, which need to create a distinct set of chemical micro-environments that are separated from each other, either in physical space or across time, but also connected. Living things achieve this by constructing systems of interacting compartments that exist at a range of scales, from the very large to the extremely small.

The biggest of these scales will probably be most familiar: the different tissues and organs of multicellular organisms like plants, and animals – like you and me. These are distinct compartments, each customized for specific chemical and physical processes. Your stomach and intestines digest the chemicals in food; your liver detoxifies chemicals and drugs; your heart uses chemical energy to pump blood, and so on. The functions of these organs all depend on the specialized cells and tissues they are made from: cells in the stomach's lining secrete acid, while those in the heart muscles contract. All those cells are, in turn, compartments in their own right.

In fact, the cell is the fundamental example of the compartmentation of life. The essential role of the cell's outer membrane is to keep the contents of the cell separate from the rest of the world. Thanks to the isolating effect of that membrane, cells can work to maintain an island of chemical and physical order. Cells can only sustain this state temporarily, of course: when they stop working, they die and chaos reasserts its grip.

The cell itself contains successive layers of compartmentation. The largest of these compartments are the membrane-bounded organelles, such as the nucleus and the mitochondria. But before we can see how these work, we need to first zoom down to the simpler level of the carbon polymers, since the bigger compartments are all built on and around the properties of these elementary components.

The smallest chemical compartments within the cell are the surfaces of the enzyme molecules themselves. To get a feeling for quite how small these molecules are, look at the very fine hairs on the back of your hand. They are among the most slender structures you can see with your naked eye, but they are massive compared to enzyme proteins. Around two thousand molecules of invertase could line up side by side across the diameter of each of those hairs.

Each enzyme protein molecule provides enclosed spaces and docking sites which have specific shapes that are tailor-made, at the scale of individual atoms, for associating with the specific molecules that they work with. These exquisite structures are far too small to see directly, even with the most powerful light-focusing microscopes. Researchers must infer their shapes and properties using techniques such as X-ray crystallography and cryo-electron microscopy, which extend our senses to an extraordinary degree, allowing us to determine the positions and properties of the hundreds or thousands of connected atoms they are made from. Researchers can then see how enzymes interact with the chemicals they manipulate during a reaction. These chemicals are called *substrates*. Enzymes and their substrates fit together like the pieces of a minute three-dimensional jigsaw puzzle. When elements of this puzzle come together, the chemical reactions are shielded from the rest of the cell and presented at just the right angle and in the right chemical conditions for the enzymes to undertake their extraordinarily precise acts of atomic surgery, manipulating individual atoms and making or breaking particular molecular bonds. Invertase, for example, works by breaking one specific bond between an oxygen atom and a carbon atom in the middle of a molecule of sucrose.

Enzymes are able to work together to ensure that the product of one reaction is passed on directly to become the substrate for the next. That way, a whole series of chemical reactions needed for complex processes, such as those needed to build lipid membranes or other complex chemical components from simpler constituents, can be co-ordinated. Biologists call these complex interacting series of chemical processes metabolic pathways, some of which involve many distinct reactions. They work rather like the assembly lines in a factory; each stage must be completed before the action moves on to the next.

Enzymes can also work together to carry out even more complex acts of synthesis, like copying DNA with extraordinary precision. The enzymes that work like this are best imagined as absolutely tiny molecular machines which are extremely accurate and reliable in their operation. Some of these molecular machines use chemical energy to do physical work in the cell. These include proteins that act as molecular 'motors', powering most movement of cells themselves and of various cargoes and structures within cells. Some function like despatch drivers, carrying cellular components and chemicals to the part of the cell where they are needed. They do this by following the complex trackways, also made from proteins, that criss-cross the cellular interior rather like an elaborately branching railway network. Researchers have made films of these minute molecular motors in action, and seen them 'walking' around the cell like tiny robots. These motors have ratchet mechanisms, that keep them moving forward, and help them avoid being knocked off course by accidental collisions with other molecules.

Versions of these molecular motors also create the forces needed to separate chromosomes and cleave dividing cells in half. And although they are each infinitesimally small, by working together in their billions, across many millions of muscle cells, these molecular motors are the things that power the wings of yellow butterflies as they flutter through our gardens, allow your eyes to follow the words on this page, and enable cheetahs to run at extraordinary speeds. Combining the tiny effects of individual proteins, working in very large numbers across many cells, leads to the real-world consequences we see all around us.

At a somewhat larger scale than individual enzymes and molecular machines, groups of proteins can dock with each other physically, to form a set of cellular devices that orchestrate more complex chemical processes. Important among these are the ribosomes, which are where proteins are made. Each ribosome is made from several dozen proteins, together with several large molecules of RNA, DNA's close chemical cousin. Ribosomes are bigger than the typical enzyme – they'd line up a few hundred, rather than several thousand, abreast across a hair's breadth, but are still far too small to see directly, without the aid of an electron microscope. Cells that are growing and reproducing have a huge demand for new proteins, so they can each contain several million ribosomes.

To build a new protein molecule, a ribosome must read the genetic code of a specific gene and translate it into the twenty-letter amino acid alphabet of proteins. To do this, the cell first makes a temporary copy of a specific gene. This copy is made from RNA. It acts as a messenger, and is indeed called messenger RNA, since it is physically transported from the genes in the nucleus to the ribosome, taking a copy of the gene's information with it. The ribosome uses the messenger RNA as a template to build the protein, by stringing together amino acids in the order dictated by the gene. By forming a separate and highly structured micro-environment, ribosomes ensure this multi-stage, multi-enzyme process occurs accurately and rapidly: it takes only about a minute for each ribosome to build an average protein, consisting of 300 or so amino acids.

Much bigger than ribosomes, although still truly minute compared to familiar human-scale objects, are the cell's organelles, each contained within their own lipid membrane wrapper. These provide the next critical layer of compartmentation in eukaryote cells. At the heart of each of these cells is the organelle we know as the nucleus. Down a microscope the nucleus is usually the most visible of the organelles. But if most cells are small – two or three of your body's white blood cells would line up across the breadth of those fine hairs on your hands – nuclei are smaller. Each one occupies only around 10% of the volume of a white blood cell. But remember, packed into that incredibly tiny space is an entire copy of all of your DNA, including all of your 22,000 genes – 2 metres of it in all when stretched out straight.

All the different chemical activity that keeps cells alive requires energy, in fact lots of energy. Today, the great majority of life forms around us ultimately derive their energy from the sun. This is what the chloroplast, another organelle critical for life, achieves. Unlike the nucleus, these don't exist in animal cells; they are found only in plants and algae. Chloroplasts are the sites of photosynthesis:

the set of chemical reactions that uses energy from sunlight to drive the transformation of water and carbon dioxide into sugar and oxygen.

The enzymes needed for photosynthesis are arranged within the two layers of membrane that surround each chloroplast. Each of the cells in the blades of grass in your local park accommodates a hundred or so of these roughly spherical organelles, all of which contain high levels of pigments called chlorophylls. These chlorophylls, which are bound to proteins, are the reason grass looks green: they absorb energy from the blue and red parts of the spectrum of light, using it to power photosynthesis, resulting in them reflecting the green wavelengths.

The plants, algae and some bacteria that can carry out photosynthesis use the simple sugars they produce as an immediate source of energy and as a raw material for building other molecules they need to survive. They also produce the sugars and carbohydrates that are consumed by so many other organisms: the fungi that feed on decaying wood, the sheep that nibble on grass, the whales that hoover up tons of photosynthesizing plankton in the sea, and all the food crops that sustain people on every continent of our world. In fact, the carbon that is so crucial for the construction of every part of our bodies comes ultimately from photosynthesis. It starts as carbon dioxide, which is drawn out of the air by the chemical reactions of photosynthesis.

The chemistry of photosynthesis has not only provided the energy and raw material to build most of the life on Earth today, it has also played a definitive role in shaping our planet's history. Life seems to have first appeared around 3.5 billion years ago, which is the age of the oldest fossils so far discovered. These were single-celled microbes, which probably derived their energy from geothermal sources. Because there was no photosynthesis during the earliest period of life on Earth, there was no major source of oxygen. As a result, there was almost no oxygen in the atmosphere, and when the planet's early life forms did encounter oxygen it would have caused them problems.

Although we think of oxygen as life-sustaining, as indeed it is, it is also a highly chemically reactive gas which can damage other chemicals, including the polymers essential for life, such as DNA. Once microbes evolved the ability to photosynthesize, they multiplied, over the millennia, to such an extent that the amount of oxygen in the atmosphere spiked. What followed, between 2 and 2.4 billion years ago, is called the Great Oxygen Catastrophe. All organisms that existed at that time were microbes, either bacteria or archaea, but some researchers think *most* of them were wiped out by the appearance of all that oxygen. It is ironic that life created conditions that nearly ended life as a whole. The minority of life forms that survived would have either retreated to places where they were less exposed to oxygen, perhaps at the bottom of the sea or deep underground, for example, or they had to adapt and evolve new chemistries needed to thrive in an oxygenated world.

Today, organisms like us humans still handle oxygen with care, but we entirely depend on it because we need it to release energy from the sugars, fats and proteins that our bodies eat, make or absorb. This is brought about by a chemical process called *cellular respiration*. The final stages of this set of reactions take place within mitochondria: another organelle compartment that is critically important for all eukaryote cells.

The principal role of mitochondria is to generate the energy that cells need to power the chemical reactions of life. That's why cells that need a lot of energy contain a lot of mitochondria: to keep your heart beating, each of the cells in the muscles of your heart must employ several thousand mitochondria. All together they occupy about 40% of the space available in those heart cells. In strictly chemical terms, cellular respiration reverses the reaction at the core of photosynthesis. Sugar and oxygen react with each other to make water and carbon dioxide, releasing a lot of energy, which is captured for later use. The mitochondria ensure this multi-step chemical reaction is highly controlled and takes place in an orderly, stepwise way, without too much energy being lost, and without reactive oxygen and electrons escaping and damaging the rest of the cell.

The key energy-capturing step in cellular

respiration is based on the movement of protons, which are single atoms of hydrogen that have been stripped of an electron to give them an electrical charge. These protons are pushed out from the centre of the mitochondrion, into the gap between the two membranes that enclose each mitochondrion. This results in the build-up of many more charged protons outside the inner mitochondrial membrane than inside. Although based on chemistry, this is essentially a physical process. You can think of it as being rather like pumping water uphill to fill a dam. In a hydroelectric power station, water from the dam is allowed to rush downhill, through turbines that turn the water's kinetic energy into electric energy. In the case of the mitochondria, protons pumped beyond the membrane 'dam' rush back into the centre of the organelle, via channels made of protein, which capture the force created by the cascade of charged particles and store it in the form of high-energy chemical bonds.

The person who first imagined that cells might produce their energy in such an unexpected way was the British biochemist and Nobel laureate

Peter Mitchell. He used to be in the Zoology Department of the University of Edinburgh, where I later worked on the yeast cell cycle, but by the time I got there, he had left to set up his own private laboratory on the moors of south-west England. This was quite an unusual thing to do, and he was considered by some to be a true British eccentric. I met him when he was in his late seventies and was impressed by his unabated curiosity and passion for knowledge. Our conversations went everywhere. I was struck by the creativity of his thinking, and impressed by the way he ignored his doubters and went on to prove that his unusual idea was, in fact, correct.

The tiny protein structures that act as the 'turbines' in the mitochondria even look a bit like the turbines in electric power stations, although they are miniaturized by a factor of several billion-fold! As protons rush through the molecular turbine, which has a channel only 10 thousandths of a millimetre wide, they turn an equally small molecular-scale rotor. That turning rotor drives the generation of an all-important chemical bond, creating a new molecule of a substance called adenosine triphosphate, or ATP for short. This happens at the rapid rate of 150 reactions per second.

ATP is life's universal energy source. Each molecule of ATP stores energy, acting like a minuscule battery. When a chemical reaction within a cell needs energy the cell breaks ATP's high-energy bond, turning ATP into adenosine diphosphate (ADP), a process that releases energy that the cell can use to trigger a chemical reaction or a physical process, such as each of the steps taken by a molecular motor.

Most of the food you eat eventually ends up being processed in your cells' mitochondria, which use the chemical energy it contains to make a prodigious quantity of ATP. To fuel all of the chemical reactions needed to support your body's trillions of cells, your mitochondria together produce, amazingly, the equivalent of your entire bodyweight in ATP every day! Feel the pulse beating in your wrist, the heat of your skin, and the rise and fall of your chest as you breathe: it's all fuelled by ATP. Life is powered by ATP.

All living things need a constant and reliable supply of energy and, ultimately, they all make their energy through the same process: controlling the flow of protons across a membrane barrier to make ATP. If there is anything remotely like a 'vital spark' that sustains life, it is perhaps this tiny flow of electric charge across a membrane. But there is nothing mystical about it: it is a well-understood physical process. Bacteria do this by actively pumping protons across their outer membrane, while the more complex cells of eukaryotes do it within a specialized compartment: the mitochondrion.

Together, all of these different levels of spatial organization within cells – from the unimaginably small docking sites within individual enzymes, to the comparatively large nucleus that contains the chromosomes – point to a new way of thinking about the cell. When we look at the beautiful and highly elaborate pictures produced by the powerful microscopes of today, we are looking at a complex and constantly changing network of organized and interconnected chemical micro-environments. This view of the cell is worlds away from that of cells as mere Lego-like building blocks for the more complex tissues and organs of plants and animals. Each cell is a complete and highly sophisticated living world in its own right.

Gradually, since Lavoisier started to ask how fermentation worked more than two centuries ago, biologists have come to recognize that even the most complex behaviours of cells and of multicellular bodies can be understood in terms of chemistry and physics. This way of thinking was very important to me and my lab colleagues as we sought to understand how the cell cycle is controlled. We had discovered the *cdc2* gene as a cell cycle controller, but we then wanted to know what the gene actually *did*. What chemical or physical processes does the Cdc2 protein it makes actually carry out?

To work this out we needed to move from the rather abstract world of genetics to the more concrete, mechanistic world of cellular chemistry. That meant we had to do biochemistry. Biochemistry tends to take a more reductionist approach, describing chemical mechanisms in great detail, whilst genetics takes a more holistic approach, looking at the behaviour of the living system as a whole. In our case, genetics and cell biology had shown us that *cdc2* was an important controller of the cell cycle, but we needed biochemistry to show *how* the protein made by the *cdc2* gene worked in molecular terms. Both approaches provide different kinds of explanations; when they agree with each other it gives you confidence that you are on the right track.

It turned out that the Cdc2 protein was an enzyme called a protein kinase. These enzymes catalyze a reaction called *phosphorylation* that adds a small phosphate molecule, which has a strong negative charge, to other proteins. For Cdc2 to function as a protein kinase it must first bind to another protein called cyclin, which activates it. Together, Cdc2 and cyclin form an active protein complex called Cyclin Dependent Kinase, or CDK for short. Cyclin was discovered and named by my friend and colleague Tim Hunt, as a protein that 'cycled' up and down in level through the cell cycle, with those changes being part of the mechanism the cell uses to ensure the CDK complex is turned 'on' and 'off' at the correct time. *Cyclin*, incidentally, is a much better name than *cdc2*!

When the active CDK complex phosphorylates other proteins, the negatively charged phosphate molecule that it adds changes the shape and chemical properties of those target proteins. That, in turn, changes the way they work. It can, for example, activate other enzymes, just as adding cyclin to the Cdc2 protein makes active CDK. Because protein kinases like CDK can rapidly phosphorylate many different proteins simultaneously, these enzymes are often used as switches in cells. That is what happens in the cell cycle. Processes such as copying the DNA in S-phase, early in the cell cycle, and separating the copied chromosomes during mitosis, late in the cell cycle, demand the co-ordinated action of many different enzymes. By phosphorylating large numbers of these different proteins all at once, CDK can exert control over complex cellular processes. Understanding protein phosphorylation is, therefore, key to understanding cell cycle control.

I cannot stress enough how satisfying it was to work all of this out and really see how *cdc2* exerted its great influence over the cell cycle. It really did feel like one of those rare eureka moments. The programme of research in my lab had moved from identifying genes in yeast, such as *cdc2*, which controlled the cell cycle and therefore cell reproduction, through to showing this control was the same in all eukaryotes from yeast to humans, to finally working out the molecular mechanism by which it acted. This took quite a long time though, a total of about fifteen years, with about ten colleagues working together in my lab. And, as is usually the case in science, it was also based on contributions from many other labs around the world, working on the cell cycle in cells from an exotic range of living organisms, including starfish, sea urchins, fruit flies, frogs, mice and, eventually, humans.

Ultimately, life emerges from the relatively simple and well-understood rules of chemical attraction and repulsion, and the making and breaking of molecular bonds. Somehow these foundational processes, operating *en masse* at a minuscule molecular scale, combine to create bacteria that can swim, lichens that grow on rocks, the flowers we tend in our gardens, flitting butterflies, and you and me, who are able to write and read these pages.

The notion that cells, and therefore living organisms, are astoundingly complicated, but ultimately comprehensible, chemical and physical machines is now the accepted way to think about life. Today, biologists build on this insight by attempting to characterize and catalogue all the components of these astonishingly complex living machines. To do this, we now have access to powerful technologies that allow deep study of the extreme complexity of living cells. We can take a cell or a group of cells and sequence all the DNA and RNA molecules they contain, and identify and count the thousands of different types of proteins present. We can also describe in detail all the fats, sugars and other molecules that are found in the cells. These technologies hugely extend the reach of our senses, giving us a new and highly comprehensive view of cells' invisible and ever-changing componentry.

Opening these new vistas onto the cell creates new challenges too. As Sydney Brenner put it: 'We are drowning in data but thirsty for knowledge.' His concern was that too many biologists spend a lot of time recording and describing the details of living chemistry, without always fully understanding what it all *means*. Central to turning all this data into useful knowledge is understanding how living things *process* information.

This is the fifth of biology's great ideas, and the one we will consider next.

5. LIFE AS INFORMATION

Working as a Whole

What was it that made that yellow butterfly venture into my childhood garden all those years ago? Was it hungry, looking for somewhere to lay its eggs, or perhaps being chased by a bird? Or was it just responding to some inbuilt urge to explore its world? Of course I do not know why that butterfly was behaving as it did, but what I can say is that it was interacting with its world and then taking action. And to do that, it had to manage information.

Information is at the centre of the butterfly's existence and indeed at the centre of all life. For living organisms to work effectively as complex, organized systems they need to constantly collect and use information about both the outer world they live in and their internal states within. When these worlds – either outer or inner – change, organisms need ways to detect those changes and respond. If they do not, their futures might turn out to be rather brief.

How does this apply to the butterfly? When it was flying about, its senses were building up a detailed picture of my garden. Its eyes were detecting light; its antennae were sampling molecules of the different chemical substances in its vicinity; and its hairs were monitoring vibrations in the air. Altogether, it was gathering a lot of *information* about the garden I was sitting in. It then brought all this diverse information together, with the aim of transforming it into useful knowledge that it could then act upon. That knowledge might have been detecting the shadow of a bird or of an inquisitive child, or recognizing the smell of nectar from a flower. This then generated an outcome: an ordered sequence of wing movements that led the butterfly to either avoid the bird or to settle on a flower to feed. The butterfly was combining many different sources of information and using them to make decisions with meaningful consequences for its future.

Closely linked with their reliance on information is the way living things act with a sense of purpose. The information the butterfly was gathering *meant* something. It was being used by the butterfly to help it decide what to do next to achieve some specific end. That meant it was acting with *purpose*.

Biology is a branch of science where it can often make sense to talk about purpose. In the physical sciences by contrast we would not ask about the purpose of a river, a comet or a gravitational wave. But it does make sense to ask that of the *cdc2* gene in yeast, or of the flight of a butterfly. All living organisms maintain and organize themselves, they grow, and they reproduce. These are purposeful behaviours that have evolved because they improve the chances of living things achieving their fundamental purpose, which is to perpetuate themselves and their progeny.

Purposeful behaviour is one of life's defining

features, but it is only possible if living systems operate as a whole. One of the first people to understand this distinctive feature of living things was the philosopher Emmanuel Kant, at the turn of the nineteenth century. In a book called *Critique of Judgement*, Kant argued that the parts of a living body exist for the sake of the whole being, and that the whole being exists for the sake of its parts. He proposed that living organisms are organized, cohesive and self-regulating entities that are in control of their own destiny.

Consider this at the level of the cell. Each cell contains a profusion of different chemical reactions and physical activities. Things would rapidly break down if all these different processes operated chaotically, or in direct competition with one another. It is only by managing information that the cell can impose order on the extreme complexity of its operations and therefore fulfil its ultimate purpose of staying alive and reproducing.

To understand how this works, remember that the cell is a chemical and physical machine that behaves as a whole. You can understand quite a

lot about a cell by studying its individual components, but to function properly, the multitude of
different chemical reactions operating within the
living cell must communicate with each other
and work together cohesively. That way, when
either its environment or its inner state changes –
perhaps the cell runs low on sugar, or encounters
a poisonous substance – it can sense that change
and adjust what it does, thereby keeping the whole
system functioning as optimally as possible. Just as
a butterfly gathers information about the world and
uses this knowledge to modify its behaviour, cells
are constantly assessing the chemical and physical circumstances both within and around them,
and using that information to regulate their own
state.

To get a better handle on what it means for cells
to use information to regulate themselves, it might
be helpful to first consider how it is achieved in more
straightforward human-designed machines. Take
the centrifugal governor, first developed for use
with millstones by the Dutch polymath Christiaan
Huygens, but adapted with great success by the

Scottish engineer and scientist James Watt in 1788. This device could be fitted to a steam engine to ensure the engine runs at a constant speed, rather than racing away and perhaps breaking down. It is comprised of two metal balls that spin around a central axis, which is powered by the steam engine itself. As the engine runs faster, centrifugal forces push the balls outwards and upwards. This has the effect of opening a valve, which releases steam from the engine's piston, slowing the steam engine down. As the engine slows, gravity pulls the steel balls of the governor back down, closing the valve and allowing the steam engine to speed up again, towards the desired speed.

We can understand Watt's governor best in terms of information. The position of the balls act as a read-out for information about the speed of the engine. If that speed exceeds the desired level, then a switch is activated – the steam valve – which reduces the speed. This creates an information-processing device which the machine can use to regulate itself, without needing any input from a human operator. Watt had built a simple mechanical device that behaves in a purposeful way. Its purpose was to keep the steam engine operating at constant speed, and it achieved that purpose brilliantly.

Systems that work in a conceptually similar way, although often through very much more complex and adjustable mechanisms, are used widely in living cells. Such mechanisms provide an efficient way of achieving homeostasis, which is the active process of maintaining conditions that are conducive to survival. It's through homeostasis that your body works to maintain a consistent temperature, fluid volume and blood sugar, for example.

Information processing permeates all aspects of life. To illustrate this, let's look at two examples of complex cellular components and processes that are best understood through the lens of information.

The first is DNA and the way its molecular structure explains heredity. The critical fact about DNA is that each gene is a *linear* sequence of information written in the four-letter language of DNA. Linear sequences are a familiar and highly effective strategy for storing and conveying information; it's the one used by the words and sentences that you are reading here, and also the one used by the programmers who wrote the code for the computer on your desk and the phone in your pocket.

These different codes all store information *digitally*. Digital here means that information is stored in different combinations of a small number of digits. The English language uses 26 basic digits, the letters of the alphabet; computers and smartphones use patterns of '1's and '0's; and DNA's digits are the four nucleotide bases. One great advantage of digital codes is that they are readily translated from one coding system into another. This is what cells do when they translate the DNA code into RNA and then into protein. In doing so, they transform genetic information into physical action, in a seamless and flexible way that no human-engineered system can yet match. And whilst computer systems must 'write' information onto a different physical medium in order to store it, the DNA molecule 'is' the information, which makes it a compact way to store data. Technologists have recognized this and are developing ways to encode information in DNA molecules to archive it in the most stable and space-efficient way possible.

DNA's other critical function, its ability to copy itself very precisely, is also a direct consequence of its molecular structure. Considered in terms of information, the molecular attractions between the pairs of bases (A to T, and G to C) provide a way to make very precise and reliable copies of the information held by the DNA molecule. This intrinsic replicability ultimately explains why information held in DNA is so stable. Some gene sequences have persisted through unbroken series of cell divisions over immense durations of time. Large parts of the genetic code needed to build the various cellular components, such as the ribosomes, for example, are recognizably the same in all organisms, be they bacteria, archaea, fungi, plants or animals. That means the core information in those genes has been preserved for probably three billion years.

This explains why the double helix structure is so important. By revealing it, Crick and Watson created a bridge linking together the geneticists' 'top down' conceptual understanding of how the information needed for life is passed down through the generations, with the 'bottom up' mechanistic understanding of how the cell is built and operated at the molecular scale. It emphasizes why the chemistry of life only makes sense when it is considered in terms of information.

The second example where information is key to understanding life is gene regulation, the set of chemical reactions cells use to turn genes 'on' and 'off'. What this provides is a way for cells to use only the specific portions of the total set of genetic information that they actually need at any given moment in time. The critical importance of being able to do this is illustrated by the development of a formless embryo into a fully formed human being. The cells in your kidney, skin and brain all contain the same total set of 22,000 genes, but gene regulation means the genes needed to make a kidney were turned 'on' in embryonic kidney cells, and those that function specifically to create skin or brain were turned 'off', and *vice versa*. Ultimately, the cells in each of your organs are different because they use very different combinations of genes. In fact, only about 4,000, or a fifth, of your total set of genes are thought to be turned on and used by all the different types of cells in your body to support the basic operations needed for their survival. The rest are only used sporadically, either because they perform specific functions only required by some types of cell, or because they are only needed at specific times.

Gene regulation also means that exactly the same set of genes can be used to create dramatically different creatures at different stages of their lives. Every elaborate and complex brimstone butterfly starts out as a rather less impressive green caterpillar; the dramatic metamorphosis from one form to the other is achieved by drawing on different portions of the same total set of information stored in the same genome and using it in different ways. But gene regulation is not only important when organisms are growing and developing, it is also one of the main ways all cells adjust their workings and structures to survive and adapt when their environments change. For example, if a bacterium encounters a new source of sugar, it will quickly turn on the genes it needs to digest that sugar. In other words, the bacterium contains a self-regulating system that automatically selects the precise genetic information it needs to improve its chances of surviving and reproducing.

Biochemists have identified many of the basic mechanisms used to achieve these various feats of gene regulation. There are proteins that function as so-called 'repressors' that turn genes off, or 'activators' that turn genes on. They do this by seeking out and binding to specific DNA sequences in the vicinity of the gene being regulated, which then makes it either more or less likely that a messenger RNA is produced and sent to a ribosome to make a protein.

It is important to know how all this works at the chemical level, but as well as asking *how* genes are regulated, we will also want to understand *which* genes are regulated, whether they're being turned on or off, and *why*. Answering these questions can lead to new levels of understanding. They can start to tell us about how the information held in the genome of a rather uniform human egg cell is used to instruct the formation of all the hundreds of different types of cell present in an entire baby; how a new heart drug can turn genes on and off to correct the behaviour of cardiac muscle cells; how we might re-engineer the genes of bacteria to make a new antibiotic; and much more besides. When we start to look at gene regulation in this way it is clear that concepts based on information processing are essential to understanding how life works.

This powerful way of thinking emerged from studies made by Jacques Monod and his colleague François Jacob; work that earned them a Nobel Prize in 1965. They knew that the *E. coli* bacteria they studied could live on one or the other of two sugars. Each sugar needed enzymes made by different genes to break it down. The question was, how did the bacteria decide how to switch between the two sugars?

These two scientists devised a brilliant series of genetic experiments that revealed the logic underlying this particular example of gene regulation. They showed that when bacteria are feeding on one sugar, a gene repressor protein switches off the key gene needed for feeding on the alternative sugar. But when the alternative sugar is available, the bacteria rapidly switch back on the repressed gene for digesting that sugar. The key to that switch is the alternative sugar itself: it binds to the repressor protein, stopping it from working properly, and thereby allowing the repressed gene to be turned back on. This is an economical and precise way of achieving purposeful behaviour. Evolution has devised a way for the bacterium to sense the presence of an alternative energy source, and to use that information to adjust its internal chemistry appropriately.

Most impressively, Jacob and Monod managed to work all of this out at a time when nobody could directly purify the specific genes and proteins involved in this process. They solved the problem by looking at their bacteria through the prism of information, which meant they did not need to know about all the specific 'nuts and bolts' of the chemicals and components underpinning the cellular process they were studying. Instead, they used an approach based on genetics, mutating genes involved in the process and treating genes as abstract informational components that controlled gene expression.

Jacob wrote a book called *The Logic of Life* and Monod wrote one called *Chance and Necessity*. Both covered similar issues to those I am discussing in this book and both greatly influenced me. I never knew Monod, but met Jacob a number of times. The last time I saw him, he invited me to lunch in Paris. He wanted to talk about his life and discuss ideas: how to define life, the philosophical implications of evolution and the contrasting contributions made by French and Anglo-Saxon scientists to the history of biology. Constantly fidgeting due to old war wounds, he was the archetypal French intellectual, incredibly well-read, philosophical, literary and political – a great and memorable meeting for me.

Jacob and Monod were working at a time when understanding was emerging of how information flowed from gene sequence to protein to cellular function, and how that flow was managed. This information-centred approach also guided my thinking. When I started my research career I wanted to know how the cell interpreted its own state and organized its internal chemistry to control the cell cycle. I did not want just to describe what happens during the cell cycle, I wanted to understand what *controlled* the cell cycle. That meant I often came back to thinking about the cell cycle in terms of information and considering the cell not only as a chemical machine but also as a logical and computational machine, as Jacob and Monod considered it – one that owes its existence and future to its ability to process and manage information.

In recent decades biologists have developed powerful tools and invested a lot of effort in identifying and counting the diverse components of living cells. For example, my lab put a lot of work into sequencing the whole genome of the fission yeast. We did this with Bart Barrell, who had worked with Fred Sanger, the person who invented the first practical and reliable way to sequence DNA back in the 1970s. I met Fred several times during this project, although he had officially retired by then. He was a rather quiet, gentle man, who liked growing roses, and, similar to many of the most successful scientists I have met over the years, always generous with his time, talking to and encouraging younger scientists. When he came to Bart's lab, he looked like a gardener who had lost his way, a gardener who had, of course, won two Nobel Prizes!

Together, Bart and I organized a collaborative effort of about a dozen labs from around Europe to read all of the approximately 14 million DNA letters in the fission yeast genome. It took about 100 people and around three years to complete, and was, if I remember correctly, the third eukaryote to be completely and accurately sequenced. That was around 2000. Now the same genome could be sequenced by a couple of people in about a day! Such have been the advances in DNA sequencing over the last two decades.

Gathering data like this is important, but only as a first step towards the crucial, and more challenging, aim of understanding how it all works together. With this objective in mind, I think most progress will be made by looking at the cell as being made up of a series of individual modules that work together to achieve life's more complex properties. I use the word module here to describe a set of components that function as a unit in order to execute a particular information-processing function.

By this definition, Watt's governor would be a 'module', one with the clearly defined purpose of controlling the speed of an engine. The gene regulatory system Jacob and Monod discovered for controlling sugar usage in bacteria is another example. In terms of information, they both work in a similar way: they are examples of information-processing modules called negative feedback loops. This kind of module can be used to maintain a steady state, and they are employed widely in biology. They work to keep your blood sugar levels relatively constant, even after you consume a sweet snack like a sugar-coated doughnut, for example. Cells in your pancreas can detect an excess of sugar in your blood and respond by releasing the hormone insulin into your bloodstream. Insulin, in turn, triggers cells in your liver, muscles and fatty tissues to absorb sugar out of your blood, reducing your blood sugar, and converting it into either insoluble glycogen or fat, which is then stored for later use.

A different type of module is the positive feedback loop, which can form irreversible switches that once turned on are never turned off. A positive feedback loop works in this way to control the way apples ripen. Ripening apple cells produce a gas called ethylene, which acts to both accelerate ripening and to increase the production of ethylene. As a result, apples can never get less ripe, and neighbouring apples can help each other to ripen more quickly.

When different modules are joined together, they can produce more sophisticated outcomes. For example, there are mechanisms that produce switches that can flip reversibly between 'on' and 'off' states, or oscillators that rhythmically and continuously pulse 'on' and 'off'. Biologists have identified oscillators that work at the level of gene activities and protein levels – these are used for

many different purposes, for example to differentiate between day and night. Plants have cells
in their leaves that use an oscillating network of
genes and proteins to measure the passing of time,
and thereby allow the plant to anticipate the
start of a new day, turning on the genes needed
for photosynthesis just before it gets light. Other
oscillators pulse on and off as a result of communication between cells. One example is the heart
that is beating in your chest right now. Another is
the oscillating circuit of neurons that ticks away
in your spinal cord, which activates the specific
pattern of repeated contractions and relaxations of
leg muscles that allows you to walk at a constant
pace. All without you having to give it any con-
scious thought.

Different modules link together in living organisms to generate more complex behaviours. A
metaphor for this is the way the different functions
of a smartphone work. Each of those functions – the
phone's ability to make calls, access the internet,
take photographs, play music, send emails and so
on – can be considered like the modules operating

in cells. An engineer designing a smartphone has to make sure all these different modules work together so the phone can do everything it needs to do. To achieve this, they create logical maps that show how information flows between the different modules. The great power of starting the design of a new phone at the level of modules is that engineers can make sure their plans make functional sense, without getting lost in the details of individual parts. That way, they need not initially give too much thought to the huge number of individual transistors, capacitors, resistors and countless other electronic components that make up each module.

Adopting the same approach provides a powerful way to comprehend cells. If we can understand the cell's different modules and see how cells link them together to manage information, we don't necessarily need to know all the minute molecular details of how each module works. The overriding ambition should be to capture *meaning*, rather than simply *catalogue* complexity. I could, for example, give you a list containing all the different words printed in this book, together with how frequently they occur. This catalogue would be like having a parts list without an instruction manual. It would give a sense of the complexity of the text, but almost all of its meaning would be lost. To grasp that meaning, you need to read the words in the correct order and develop an understanding of how they convey information at higher levels, in the form of sentences, paragraphs and chapters. These work together to tell stories, give accounts, connect ideas and make explanations. Exactly the same is true when a biologist catalogues all the genes, proteins or lipids in a cell. It is an important starting point, but what we really want is an understanding of how those parts work together to form the modules that keep the cell alive and able to reproduce.

Analogies derived from electronics and computing, like the smartphone example I used just now, are helpful in understanding cells and organisms, but we must use them with care. The information-processing modules used by living things and those used in human-made electronic circuitries are in some respects very different.

Digital computer hardware is generally static and inflexible, which is why we call it 'hardware'. By contrast, the 'wiring' of cells and organisms is fluid and dynamic because it is based on biochemicals that can diffuse through water in the cells, moving between different cellular compartments and also between cells. Components can be reconnected, repositioned and repurposed much more freely in a cell, effectively 'rewiring' the whole system. Soon, our helpful hardware and software metaphors begin to break down, which is why the systems biologist Dennis Bray coined the insightful term 'wetware' to describe the more flexible computational material of life. Cells create connections between their different components through the medium of wet chemistry.

This is also true in the brain, the archetypal and highly complex biological computer. Throughout your life, nerve cells are growing, retracting and making and breaking connections with other nerve cells.

For any complex system to behave as a purposeful whole, there needs to be effective communication between both the different components of the system and with the outside environment. In biology, we call the set of modules that carry out this communication signalling pathways. Hormones released into your blood, like the insulin that regulates your blood sugar, are one example of a signalling pathway, but there are many others too. Signalling pathways transmit information within cells, between cells, between organs, between whole organisms, between populations of organisms and even between different species across whole ecosystems.

The way signalling pathways transmit information can be adjusted to achieve many different outcomes. They can send signals that simply turn an output on or off, like a light switch, but signals can work in more subtle ways too. In some situations, for example, a weak signal switches on one output and a stronger signal switches on a second output. In a similar way a whisper gets your immediate neighbour's attention, but a shout is needed to evacuate a whole room in an emergency. Cells can also exploit the dynamic behaviour of signalling pathways to transmit a far richer stream of information. Even if the signal itself can only be 'on' or 'off', more information can be transmitted by varying the time spent in each of those two states. A good analogy is Morse code. Through simple variations in the duration and order of signal pulses, the 'dots' and 'dashes' of Morse code can convey streams of information that overflow with meaning, be it an SOS call or the text of Darwin's *On the Origin of Species*. Biological signalling pathways that behave in this way can generate information-rich properties that carry more meaning than signalling sequences conveying a simple 'yes/no' or 'on/off' message.

As well as signalling through space, cells need ways to signal through time. To achieve this, biological systems must be able to store information. This means that cells can carry with them chemical imprints of their past experiences, which we can think of as working a bit like the memories we form in our brains. These cellular memories range widely, from transient impressions of what happened just a moment ago, to the extremely long-term and stable memories held by DNA. The cell uses short-term historical information during the cell cycle, when the status of events that occur early in the cycle are 'remembered' and signalled forward to later events in the cycle. For example, if the process of copying DNA has not yet been completed or has gone wrong, this fact needs to be registered and relayed to the mechanisms which bring about cell division. If not, the cell could attempt to divide before its entire genome has been properly copied, which could result in the loss of genetic information and the death of the cell.

The processes involved in gene regulation allow cells to store information over longer time scales. This was a particular interest of the British biologist Conrad Waddington during the mid-twentieth century. I met Waddington at Edinburgh University when I started my postdoctoral research there in 1974. He was a striking character, with wide interests in art, poetry and left-wing politics, but he is best known for coining the word *epigenetics*. He used it to describe the way cells gradually take on more specialized roles during the development of an embryo. Once the growing embryo instructs cells to commit to these roles, they remember that information and rarely change track. That way, once a cell has committed to forming part of the kidney, it will remain part of the kidney.

Today, the way most biologists use the word epigenetics is based on Waddington's ideas. It describes the set of chemical reactions that cells use to turn genes either on or off in fairly enduring ways. These epigenetic processes do not change the DNA sequence of the genes themselves; instead, they often work by adding chemical 'tags' to the DNA, or to proteins that bind to that DNA. This creates patterns of gene activity that can persist through the lifespan of a cell and sometimes even longer, through many cell divisions. Occasionally, although far less commonly, they can persist from one generation to the next, potentially carrying information about an individual organism's life history and experience directly, in chemical form, from parents to their offspring and on to subsequent generations. Some have argued that the cross-generational persistence of these patterns of gene expression poses a major challenge to the idea that inheritance is based only on the DNA sequences encoded in genes. However, present evidence indicates that cross-generational epigenetic inheritance only occurs in a few instances and seems to be very rare in humans and other mammals.

In addition to gene regulation, information processing is important for the ways living beings create ordered structures in space. Take my brimstone butterfly. It is an exquisitely complex construction: the wings are carefully shaped to allow it to fly; there are spots and veins placed on those wings with great precision. Moreover, all individual butterflies are built to the same plan: they all have a head, thorax and abdomen, six legs and two antennae, for example. These structures all form and grow in the same predictable proportion to the rest of their bodies. How is all this extraordinary spatial structure generated? How does it all emerge from a single uniform egg cell?

Even cells can take on a range of highly elaborate structures and shapes that are quite distinct from the regular, box-like cork cells that Robert Hooke described in the seventeenth century and that I observed in onion roots as a schoolboy – there are the comb-like hairs on lung cells, whose constant beating pushes mucus and infections out of your lungs; cube-shaped cells that live in and manufacture your bones; and neurons whose long, branching connections reach all parts of your body; among very many others. And within those cells, their organelles can be precisely located and grow and adjust their position as the cell changes.

How all of this spatial order develops is one of the more challenging questions in biology. Satisfactory answers will depend on understanding how information is signalled through both space and time. At present, we only really understand fully the structure of biological objects that are direct assemblies of molecules. The ribosome is a good example. The shapes of these relatively small objects are determined by the chemical bonds that form between their molecular components. You can think of these structures as if they are built up by adding pieces to a three-dimensional jigsaw, a bit like Lego. That means the information needed to assemble these structures is embodied in the shape of the ribosome components themselves – the proteins and RNAs. Those shapes, in turn, are ultimately specified very precisely by the information held in the genes.

Understanding how structures form at larger scales, in objects such as organelles, cells, organs and whole organisms is more difficult. Direct molecular interactions between components cannot explain how these structures form. That's partly because they are larger, sometimes much larger, than objects like ribosomes. But it is also because they can produce and maintain perfect structures over a range of different sizes, even when cells or bodies grow or shrink. That is simply not possible with fixed, Lego-like molecular inter-actions. Take the division of a cell for example. A cell has a well-organized overall structure, and when the cell divides it generates two cells of approximately half the size and yet each of them has the same overall structure as the original 'mother' cell.

A similar phenomenon is seen with the development of an embryo, such as a sea urchin. A fertilized sea urchin egg undergoes repeated cell divisions and generates an elaborate and rather beautiful little organism. If the two cells formed after the very first division of the egg are split apart, then each cell will generate two perfectly formed sea urchins, but, amazingly, each one will be just half the size of a normal urchin of that age. This self-regulation of size and form is extraordinary and has puzzled biologists for more than a century.

However, by thinking about information, biologists are beginning to make sense of how these things take shape. One way that developing embryos generate the information they need to transform a uniform cell or group of cells into a highly patterned structure is by making chemical gradients. If you put a small drop of ink into a bowl of water, it will slowly diffuse away from the location of the original drop. The intensity of the ink colour gets lower further away from the drop, making a chemical gradient. That gradient can be used as a source of information: for example, if the concentration of ink molecules is high, we know we are close to the centre of the bowl, where the ink was dripped in.

Let's now replace the bowl with a ball of identical cells and, instead of ink, we inject one side of the ball with a dose of a particular protein that can change the properties of cells. What this provides is a way to add spatial information to those cells so they can begin to build a pattern. The protein will diffuse through the cells, forming a gradient of high concentration at one side of the ball and low concentration on the other side. If cells react differently to high and low concentrations, the protein gradient can provide the information needed to start constructing a complex embryo. If, for example, a high protein concentration made head cells, a medium concentration made thorax cells, and a low concentration made abdomen cells, then one simple protein gradient could, in principle, lead to the beginnings of a new brimstone butterfly. In life, things are usually not quite as simple as that, but there is good evidence that gradients of signalling molecules across the bodies of developing organisms do indeed contribute to the appearance of sophisticated biological forms.

This was a set of problems that Alan Turing – he of Enigma code-cracking fame and one of the founders of modern computing – turned to during the early 1950s. He came up with an alternative, and imaginative, suggestion for how embryos generate spatial information from within. He devised a set of mathematical equations that predicted the behaviour of chemical substances interacting with each other, and so undergoing specific chemical reactions as they diffuse through a structure. Unexpectedly, his equations, which he called reaction-diffusion models, could arrange chemical substances into elaborate and often rather beautiful spatial patterns. By tweaking the parameters of his equations, the two substances could organize themselves into evenly spaced spots, stripes or blotches, for example. The attractive thing about Turing's model is that the patterns emerge spontaneously, according to relatively simple chemical rules of interaction between the two substances. In other words, this provides a way for a developing cell or organism to generate the information it needs to take shape, entirely from within; it is self-organizing. Turing died before his theoretical ideas could be tested in real embryos, but developmental biologists now believe that this could be the mechanism that puts spots on cheetah's backs and stripes on many fish; distributes the hair follicles on your head; and even divides each of a developing human baby's hands into five distinct fingers.

When we look at life in terms of information, it is important to appreciate that biological systems have evolved gradually over many millions of years. As we have seen, life's innovations arise as a consequence of random genetic mutations and variations. These are then sifted by natural selection, with those that work well being assimilated into the surviving, more successful, living organisms. This means that existing systems are changed progressively, by the gradual accretion of 'add-ons'. This is in some ways analogous to your phone or computer, which frequently require the downloading and installation of new software updates.

The devices gain new functions, but the software that drives them also becomes steadily more complicated. Similarly for life, all of these genetic 'updates' mean that the whole system of the cell will gradually tend to become more complex with time. This can lead to redundancy: some components will have overlapping functions; others will be the relics of superseded parts; and some will be wholly unnecessary for normal functioning but might be able to compensate if the primary component breaks.

This all means that living systems are often less efficient and rationally constructed than control circuits designed intelligently by human beings, another reason why analogies between biology and computing can only go so far. As Sydney Brenner observed, 'Mathematics is the art of the perfect. Physics is the art of the optimal. Biology, because of evolution, is the art of the satisfactory.' The life forms that survive natural selection persist because they *work*, not necessarily because they do things in the most efficient or straightforward way possible. All this complexity and redundancy makes the analysis of biological signalling networks and information flow challenging. Very often Occam's razor – looking for the simplest adequate explanation to explain a phenomenon – simply does not apply. This can disturb some physicists who turn their attentions to biology. Physicists tend to be attracted to elegant, simple solutions, and can be less comfortable with the messy and less-than-perfect reality of living systems.

My lab has frequently wrestled with the redundancies and intricacies brought about by natural selection, because they can obscure the core principles of how biological processes work. To tackle this we genetically engineered yeast cells to generate a much simplified cell cycle control circuit. It was like stripping a car of all the components that are not essential for its critical functions, such as the bodywork, the lights and the seats, leaving only the essentials – the engine, transmission and wheels. This worked better than I had hoped. Our simplified cells could still carry out the major aspects of cell cycle control. Stripping a complex mechanism down to its basic elements made it easier for us to analyse information flow, and there-fore gain new insights into the cell cycle control system.

Among the select group of indispensable cell cycle regulators highlighted by this experiment was the *cdc2* gene. As a yeast cell moves through the cell cycle, the cell itself grows steadily and the amount of the Cdc2- and cyclin-containing CDK protein complex increases too. In terms of information, the cell uses the amount of active CDK complex present as both an input that reflects information about the size of the cell, and as the crucial signal that triggers the major events of the cell cycle. Proteins required early in the cell cycle are phosphorylated by the CDK complex early, leading to the copying of DNA during S-phase, and those required later are phosphory-lated later, leading to mitosis and cell division at the end of the cell cycle. The 'early' proteins are more sensitive to the CDK enzyme activity than the 'late' ones, so they will be phosphorylated when there is less CDK activity in the cell.

This simple model of cell cycle control identified CDK activity as the crucial co-ordinating hub at the centre of cell cycle control. The explanation had just been obscured from our view by the superficial complexities of the network, the redundant functions of different components, the presence of less important control mechanisms, and perhaps also by the tendency of the human mind to embrace complexity, rather than seek out simplicity.

For much of this chapter I have focused on cells because they are the basic units of life, but the implications of thinking about life as information extend beyond the cell. There is real potential to gain powerful new insights into all parts of biology by looking for ways to understand how molecular interactions, enzyme activities and physical mechanisms produce, transmit, receive, store and process information. As this becomes a more prevalent approach, it is possible that biology will shift away from the rather common-sense and familiar world that it has generally occupied in the past, to one that is more abstract. In this, it might parallel the great shifts that took place in physics, from

Isaac Newton's essentially common-sense world to Albert Einstein's universe, ruled by relativity, and on further to the quantum 'weirdness' revealed by Werner Heisenberg, as well as Erwin Schrödinger, in the first half of the twentieth century. It might be that the complexity of biology will lead to strange and non-intuitive explanations, and to work these out biologists will need ever more assistance from scientists in other disciplines, such as mathematicians, computer scientists and physicists – even philosophers, who are more used to thinking abstractly and are less focused on our everyday experiences of the world.

A view of life that is centred on information will also help us understand higher levels of biological organization. It can shed light on how cells interact with each other to generate tissues, how tissues make organs, and how organs work together to produce a fully operational living organism, such as a human being. The same is true at even bigger scales, when we look at how living organisms interact with each other, both within species and between species, and how ecosystems and the biosphere operate. The fact that information management occurs at all scales, from the molecular to the planetary biosphere, has important implications for how biologists try to make sense of life's processes. Often, it is best to seek explanations close to the level of the phenomenon being studied. To be satisfactory, those explanations do not always need to be reduced down to the molecular-scale realm of genes and proteins.

However, it may well be that there are commonalities between the way information is managed at one scale that can illuminate how things work in a system that is either larger or smaller. For example, the logic underpinning feedback modules that control metabolic enzymes, regulate genes or maintain bodily homeostasis, will have similarities with the feedback modules that allow ecologists to make better predictions about how natural environments are likely to change when specific species go extinct or migrate out of their traditional ranges as a result of climate change or habitat destruction.

Given my interest in beetles and butterflies and insects in general, I am increasingly worried about the falling numbers and diversity of insects that are being observed in many parts of the world. What is particularly disturbing is that we do not know why this is happening. Is it habitat destruction, climate change, agricultural monocultures, light pollution, overuse of insecticides, or something else? There are many explanations proposed and some people feel very certain of their particular theories, but the truth is we do not really know. If we are to do something to help reverse declining insect populations, we need to understand the interactions between them and the rest of their worlds. This will be greatly informed by scientists who work in different ways, collaborating and thinking about these issues in terms of information.

Whichever level of biological organization we look at, attempts to deepen our comprehension will hinge on our ability to understand how information is managed within them. It is a way to move from *describing* complexity to *understanding* complexity. Once we can do this, we can start to see how flitting butterflies, sugar-consuming bacteria, developing embryos and all other life forms make the crucial leap of transforming information into meaningful knowledge that they can use to fulfil their purpose of surviving, growing, reproducing and evolving.

From our advancing understanding of the chemical and informational foundations of life springs the growing ability not only to comprehend life, but also to intervene in the workings of living things. So before I use the insights we have gained from climbing our five steps to define what life is, I want to consider how we can use knowledge of biology to change the world.

CHANGING THE WORLD

In 2012, I was due to travel to the Antarctic research station at Scott Base. I had always wanted to visit the vast frozen desert of the South Polar regions – literally the end of the Earth – and finally I had my chance. Before the trip I had to have a routine medical check-up, but the results turned out to be far from routine. For the first time in my life I had to directly confront my own mortality.

I had serious heart disease. Within a couple of weeks of this unwelcome revelation I was anaesthetized and laid out in an operating theatre. The surgeon opened my chest and identified the defective blood vessels that were failing to supply the muscles of my heart with enough blood. He then harvested four short lengths of an artery from my chest and a vein from my leg, and plumbed them into my heart in such a way that blood could bypass the problem areas. A few hours later, I woke up, battered and bruised, but with a repaired heart.

That operation saved my life. As well as the great skill and compassion of the medical staff who treated me, the success of the operation rested entirely on our understanding of what life is. Every step was guided by knowledge of the human body and the tissues, cells and chemistry within. The anaesthetist was confident that the drugs they delivered would make my brain lose consciousness in a reversible way. A solution was infused into my heart, which completely stopped it beating for some hours. It contained potassium at a concentration the doctors knew was just high enough to alter the chemistry of my heart muscle cells to make them relax. A machine stood in for my heart and lungs, oxygenating my blood properly and delivering it at the correct rate. During and after the operation, I was given antibiotics to keep infectious bacteria at bay. Without all that accumulated knowledge about life, the chances are I would not be writing these words today.

As our understanding of life has grown we have acquired great new powers to manipulate and change living things. But we must wield these powers properly. Living systems are complicated, so if we interfere with them before we understand them well enough, we will get it wrong and could cause more problems than we solve.

Throughout history, most human lives have been ended not by old age, but by infectious diseases. The attacks made by bacteria, viruses, fungi, worms and a host of other parasites and pestilences have claimed countless millions of lives, many of them before they leave infancy. The bubonic plague that swept around the world in the four-teenth century killed nearly half of all people in Europe. For much of history, death was a constant presence in daily life.

That is not quite so true today. Where vaccines, sanitation and antimicrobial drugs are available, we have the tools we need to prevent, treat or contain a wide range of once-deadly infectious diseases. Even HIV, once billed by some as the next great plague, can now, with the right care, be treated as a stable chronic condition. After millennia in which healthcare relied chiefly upon superstition, vague explanations and a host of unproven and sometimes risky remedies, this transition is a truly miraculous change. It all rests upon our knowledge of life, generated by science, and then applied to the world.

However, there is still a long way to go in fighting the ancient scourge of infectious disease. As I write these words in spring 2020, the coronavirus pandemic is spreading turmoil across the world. Like the disease caused by this coronavirus, COVID-19, many viral infections can be immobilizing, or even lethal. And although the outbreak of Ebola that ripped through West Africa in 2014–15 inspired the impressively rapid development of effective vaccines, such interventions are only helpful when they can get to the people who need them at the right time. In rich and poor countries alike, too many populations still lack good access to proven treatments. It is also astonishing that politicians in some developed nations should have ignored advice from scientists and experts and have weakened measures to deal with epidemics and pandemics such as these. This neglect has already led to grave consequences. Putting all this right should be an urgent priority for humanity.

Those of us lucky enough to live in societies that provide good medical care should cherish the protection it affords us. It is a mark of a civilized society that medical care such as the heart surgery I received from the UK's National Health Service is free at the point of delivery, regardless of the patient's ability to pay. 'Pay as you go' healthcare systems punish the poorest, and risk-based insurance systems punish the most needy. And then there are those who wilfully criticize the safety and effectiveness of vaccines without adequate evidence. They should remember that rejecting proven, clinically approved vaccines is a question of morality. By doing so they are not just imperilling the safety of themselves and their families, but also that of many others around them by disrupting herd immunity and allowing infectious diseases to spread more easily.

The battle with infectious disease is one that we will never wholly win, however. That's because of evolution by natural selection. Since most bacteria and viruses can reproduce very quickly, their genes can also adapt rapidly. This means new strains of disease can emerge at any moment, and they are constantly evolving ingenious ways to elude or trick our immune systems and medicines. That's why the rise of antimicrobial resistance is such a threat. It is natural selection in action and it is taking place before our eyes, with alarming consequences. Exposing bacteria to antibiotics without actually killing them off entirely, makes it more likely that they will evolve resistance to the drugs. That's why it is important to take the right dose of antibiotics – and only when truly needed – and to finish the course of treatment you are prescribed. Not doing so may not only put your health at risk, but also that of other people. Just as dangerous, or even more so, are the farming systems that drip-feed low doses of these drugs to animals to make them grow faster.

We are now seeing the emergence of strains of bacteria that can resist every intervention we can make; the diseases they cause are becoming untreatable. Resistant bacteria like this could cast medicine back in time, putting millions of lives at risk. Imagine a world where you or your family could be struck down by an incurable infection because of a scratch from a rose thorn, a nip from a dog or even a visit to a hospital. But we must not be fatalistic about this threat. Identifying a problem is the crucial first step towards solving it. We can and must use the antimicrobial drugs we do have more carefully; we can also design better ways to detect and track drug-resistant infections; and we need to develop potent new antimicrobial drugs and make sure the researchers who do this are well supported. We must use all our knowledge about life to solve this problem – our future may depend on it.

As healthcare has improved and the threat posed by infectious diseases has been gradually pushed back, average life expectancy has crept steadily upwards. But as people live longer, they have had to confront a host of unpleasant non-infectious conditions of ill health, including heart disease, diabetes, a range of mental health conditions and cancer. Their fundamental causes are old age and unhealthy lifestyles. Globally, they are all on the rise, and they create big challenges for both sufferers and the scientists who want to understand and treat them.

Consider cancer – it is actually not one disease, but many. Every cancer is different, and each incidence changes with time, so that an advanced cancer is often a bit like an ecosystem in its own right, containing many different types of cancer cell, each containing different genetic mutations. Once again, this is the work of evolution by natural selection. Cancers begin when cells acquire new genetic changes and mutations that cause them to start dividing and growing in uncontrolled ways. They flourish because they have a selective advantage: they can monopolize the body's resources, grow more than the non-mutated cells around them, and ignore the body's 'stop' signals.

Some of the most promising new approaches to cancer therapy have been informed by our

improved understanding of life. Cancer immuno-
therapies, for example, seek to educate the body's
immune system to recognize and attack cancer
cells. This is a smart approach because the immune
system can launch extremely precise assaults on
cancer cells, while ignoring healthy cells nearby.
New treatments are also emerging from work that
my colleagues and I started on the cell cycle of
the lowly yeasts. Drugs that bind to and inacti-
vate human versions of the CDK cell cycle control
proteins are now used to treat many women with
breast cancer. Four decades ago, I had no idea that
work on the cells of yeast would eventually lead so
directly to new cancer treatments. Because cancer
is an inevitable result of the cell's capacity to adapt
and evolve, we will never entirely eliminate it. But
as our understanding of life gets better we will
increasingly be able to spot cancer early and treat
it more effectively. I am confident there will come
a time when cancer no longer arouses fear, as it still
does today.

If we want to accelerate progress in tackling
cancer and other non-infectious diseases, decoding

the information in our genes provides important new ways forward. When the first draft DNA sequence of the human genome was shared with the world in 2003 it promised to open the door to a new future of preventative medicine. Many of those involved looked forward to a world where any individual's genetic risk factors could be accurately calculated at the moment of birth, including predictions about how those risks would interact with lifestyle and diet. But realizing this aim is very challenging, both scientifically and ethically.

That's partly because of life's profound complexity. Few human characteristics behave like the clear-cut characters of the pea seedlings that Mendel studied in his garden. The diseases that are caused in a similar way, by defective versions of single genes, include Huntington's disease, cystic fibrosis and haemophilia. Together these diseases cause a great deal of suffering and pain, but they each affect relatively small numbers of people. Most common diseases and disorders, including heart disease, cancer and Alzheimer's disease, by contrast, have more multi-factorial triggers. They are caused by the combined influences of many individual genes which operate and interact with each other and with the environments we live in in complicated and hard-to-predict ways. We are starting to unravel the intricate chains of cause and effect that intertwine our nature and our nurture, but all progress is hard-won and slow.

This is an area where understanding Life as Information comes to the fore. Researchers are now amassing extremely large collections of data – some of them containing gene sequences, lifestyle information and medical records gathered from up to millions of different people. But making sense of such large data sets is difficult. The interactions between genes and environment are so complex that the researchers who study them are stretching the limits of presently available techniques, including new approaches such as machine learning.

Useful insights are emerging, though. It is now possible to use genetic profiling to identify people with an elevated risk of suffering heart disease, or becoming obese, for example. These can be used to give advice about lifestyle and drug treatments that is tailored to individuals. This is good progress, but as the ability to make accurate predictions from our genomes gets better, we must think hard about how this knowledge should best be used.

Accurate genetic predictors of ill health pose particular difficulties for medical systems that are funded by personal health insurance. Without strict controls on how gene information can be used, individuals could find themselves being deemed uninsurable and denied care, or charged unaffordably high insurance premiums through no fault of their own. There are no such problems with medical systems that provide care that is free at the point of delivery, since they will be able to use advances in genetic predisposition to predict, diagnose and treat disease more easily. That said, this is not always easy knowledge to live with. If genetic science advanced to the point where it could make a reasonably accurate prediction of when and how you are most likely to die, would you want to know?

Then there is deciphering the genetic factors that influence non-medical factors, such as general intelligence and educational attainment. As we learn more about genetic differences between individuals, genders and populations, we must make sure these insights are never used as the basis for discrimination.

Advancing in parallel with the ability to read genomes is the ability to edit and rewrite them. An enzyme called CRISPR-Cas9 is a powerful tool that functions like a pair of molecular scissors. Scientists can use it to make very precise cuts in DNA, in order to add, delete or alter gene sequences. This is what is referred to as gene-editing, or genome-editing. Biologists have been able to do this in simple organisms, such as yeast, since around 1980, which is one of the reasons that I have worked with fission yeast, but CRISPR-Cas9 vastly improves the speed, accuracy and efficiency with which DNA sequences can be edited. It also makes it much easier to edit the genes of many more species, including human beings.

In time, we can expect new therapies based on gene-edited cells. Researchers are already making cells that are resistant to specific infections, such as HIV, or using them to attack cancers, for example. But for the time being, it is extremely reckless to attempt to edit the DNA of early stage human embryos, which would result in genetic changes in all the cells of the person born, and those of any children they might have in the future. At present there is a risk that gene-based therapies might accidentally change other genes in the genome. However, even if only the desired gene is edited, those genetic changes could also cause hard-to-predict and potentially dangerous side effects. We simply don't yet understand our genomes well enough to know for sure. There may come a time when this procedure is deemed safe enough to free families from certain genetic diseases, like Huntington's or cystic fibrosis. But using it for more cosmetic purposes, like creating babies with enhanced intelligence, great beauty, or high athletic ability is another matter altogether. This area entails one of the most thorny of today's ethical concerns about the application of biology to human life. But although talk of using gene editing to make designer babies is mostly hot air at present, many parents-to-be will have to contemplate some challenging issues in the years and decades to come, as scientists develop more powerful abilities to predict genetic influences, modify genes and manipulate human embryos and cells. All these issues need to be discussed by society as a whole, and they need to be discussed now.

At the other end of life, advances and developments in cell biology are providing ways to treat degenerative diseases. Take stem cells, for example: these are cells that the body maintains in an immature state, rather like those present in an early embryo. The key property of stem cells is their ability to divide repeatedly, to produce new cells that can then go on to adopt more specialized properties. A growing fetus or a baby contains large numbers of stem cells, since they have a constant requirement for new cells. But stem cells also persist in many different parts of the adult body long after it has stopped growing. Many millions of your body's cells die or are shed every day. That's why your skin, your muscles, the lining of your gut, the edges of the corneas in your eyes, and many other tissues of your body contain populations of stem cells.

In recent years scientists have worked out how to isolate and culture stem cells and to push them to develop into specific cell types – nerve, liver or muscle cells, for example. It is also now possible to take fully mature cells from a patient's skin and treat them in such a way that they turn back the developmental clock, reverting to a stem cell state. This raises the exciting prospect that it might one day be possible to take a swab from inside your cheek and use the cells to generate almost any other cell in your body. If scientists and doctors can fully master these techniques, and can establish that they are safe, they could potentially revolutionize treatment of degenerative disease and injuries and revolutionize transplant surgery. It might even become possible to reverse currently incurable conditions of the nervous systems and muscles, like Parkinson's disease or muscular dystrophy.

This progress is part of what has inspired bold predictions, many of them emanating from firms based in Silicon Valley, of a fast-approaching future in which it will be possible to arrest or even *reverse* ageing. It is important to keep these claims grounded in practical reality. Personally, I will not be opting to cryopreserve my brain or body when my time is up, in anticipation of a highly unlikely future time when I might be resuscitated, rejuvenated and kept alive into perpetuity. Ageing is the end product of the combined damage, death and pre-programmed shutdown of a body's cells and organ systems. Even for those in fine health, skin becomes less elastic, muscles lose tone, the immune systems becomes less responsive, and the power of the heart gradually weakens. There is no single cause for all this and it is, therefore, very unlikely that there can be a straightforward fix. But I have little doubt the decades ahead will see life expectancy creeping on upwards and – importantly – the quality of life improving in old age. We will not live for ever, but we could all benefit from ever more refined treatments that use combinations of stem cells, novel drugs and gene-based therapies, as well as healthy lifestyle practices, to revive and regenerate many parts of elderly and ailing bodies.

The application of biological knowledge

has not only revolutionized our ability to mend broken bodies, it has allowed humankind as a whole to flourish. Beginning around 10,000 BCE, the world's population leaped upwards when our ancestors started farming. They didn't see it this way at the time, but this was achieved by our human ancestors applying the principles of artificial selection to domesticate animals and plants. A much larger and more reliable supply of food was the reward.

Compared to the prehistorical surge, the world's population has grown even more dramatically within my lifetime: it has nearly tripled since I was born in 1949. That means nearly 5 billion extra mouths that must be fed each day, with all that extra food being produced on roughly the same area of agricultural land. The Green Revolution, which started in the 1950s and 60s, was key to making this possible. This involved developments in irrigation, fertilizers, pest control and, most importantly, the creation of new strains of staple food crops. In contrast to breeders throughout history, the scientists involved were able to leverage all that had been gleaned about genetics, biochemistry, botany and evolution towards the production of novel plant varieties. It was astonishingly successful and generated new crops with significantly higher yields. This hasn't been an entirely cost-free exercise, however. Some of today's intensive farming practices have a damaging effect on the land, farmers' livelihoods, and on other species that share the environment with food crops. The amount of food wasted every day is a scandal that must also be solved. But without that major injection of biological knowledge into farming practice last century, millions more would starve every year.

The global population continues to grow today. As it does, there is increasing concern about the damage human activity does to the living world. Looking ahead, we face the stark combined challenge of eking out yet more food from the land, whilst also trying to reduce our environmental impact. I think we will need to go beyond the methods that drove last century's agricultural revolution and devise even more efficient and creative ways of producing food.

But unfortunately, since the 1990s, attempts to create genetically modified (GM) strains of plants and livestock with enhanced properties have often been blocked. Frequently this has had little to do with scientific evidence and understanding. I have seen debates about the safety of GM foods constantly derailed by misunderstanding, misplaced lobbying and the injection of misleading information. Consider the case of golden rice, which has been genetically engineered to incorporate a bacterial gene into one of the rice plant's chromosomes, which makes it produce large quantities of vitamin A. There are an estimated 250 million preschool children across the world who are deficient in vitamin A, which is a significant cause of blindness and death. Golden rice might provide a direct way to help, yet it has been attacked repeatedly by environmental campaigners and non-governmental organizations (NGOs) who have even vandalized field trials set up specifically to test its safety and its effect on the environment.

Is it really acceptable to deny the world's poorest access to inventions that could help their health and food security, especially if that denial is based on fashion and ill-informed opinion rather than sound science? There is nothing intrinsically dangerous or poisonous about foodstuffs made using GM methods. What really matters is that *all* plants and livestock should be similarly tested for their safety, efficiency and predicted environmental and economic impact, regardless of how they have been made. We need to consider what the science has to say about risks and benefits, uncoloured by either commercial interests of companies, the ideological opinions of NGOs, or the financial concerns of both.

In the coming decades, I think that we will have to use genetic engineering techniques more. This could be an area where the relatively new branch of science known as *synthetic biology* could make an impact. Synthetic biologists seek to go beyond the more focused and incremental approaches traditionally used in genetic engineering, to write more radical changes to organisms' genetic programming.

The technical challenges here are substantial,

and there are questions about how we control and contain these new species, but the potential rewards could be significant. That's because life's chemistry is far more adaptable and efficient than most chemical processes people have been able to carry out in labs or factories. With GM and synthetic biology we could reorganize and repurpose life's chemical brilliance in powerful new ways. It should be possible to use synthetic biology to create nutritionally enhanced crops and livestock, but it could be applied more broadly than that. It could see us creating re-engineered plants, animals and microbes that produce entirely new types of pharmaceuticals, fuels, fabrics and building materials.

Novel engineered biological systems might even help tackle climate change. The scientific consensus is clear that our planet has entered a phase of accelerating global warming. This is a grave threat to our future and to that of the wider biosphere that we are but one part of. An increasingly urgent challenge is to reduce the amount of greenhouse gases that we emit and reduce the extent of warming. If we could re-engineer plants to carry out photosynthesis even more efficiently than they do, or make it work at an industrial scale, outside the confines of living cells, it might be possible to make biological fuels and industrial feedstocks that are carbon neutral. Scientists may also be able to engineer novel plant varieties that can thrive in marginal environments, for example in degraded soils or areas that are prone to drought, that previously have not supported cultivation. Such plants could be used not only to feed the world but also to draw down and store carbon dioxide to help manage climate change. They could also form the basis of living factories that work in sustainable ways. Instead of relying on fossil fuels, it might be possible to produce biological systems that will feed more effectively off waste, by-products and sunlight.

In parallel with these engineered life forms, another goal would be to increase the total area of the planet's surface that is covered by naturally occurring photosynthesizing organisms. This is not such a straightforward proposal as it might first seem. To make a meaningful impact it needs to be implemented at a massive scale, and also there needs to be consideration of the issue of long-term carbon storage once the plants have died or been harvested. It could involve more forests, cultivating algae and seaweed in the oceans, and encouraging the formation of peat bogs. But making any intervention work effectively and quickly enough will stretch our understanding of ecological dynamics to its limits. The ongoing, widespread, and largely unexplained decline in insect numbers is a case in point. Our future is tied to insect species, since they pollinate many of our food crops, build soils, and more besides.

Progress in all these applications requires better understanding of life and how it works. Biologists of all disciplines – molecular and cellular biologists, geneticists, botanists, zoologists, ecologists and beyond – all need to work alongside one another to help ensure human civilization continues to flourish, together with, rather than at the expense of, the rest of the biosphere. For any of this to succeed we need to face up to the scale of our ignorance. Despite the great progress we have made in understanding how life works, our present understanding is partial, sometimes very much so. If we want to interfere with living systems constructively – and safely – to achieve some of our more ambitious practical goals, we still have much to learn.

The development of new applications should always move forward hand in hand with efforts to learn more about how life works. As the Nobel prizewinning chemist George Porter once put it: 'To feed applied science by starving basic science is like economizing on the foundations of a building so that it may be built higher. It is only a matter of time before the whole edifice crumbles.' But by the same token it is self-indulgent of scientists not to recognize that useful applications should be generated wherever possible. When we see opportunities to use that knowledge for the public good we must do so.

This creates fresh questions and further problems, however. How do we agree on what we mean by the 'public good'? If new cancer therapies are hugely expensive, who should get them and who should not? Should advocating vaccine refusal without adequate evidence, or the misuse of antibiotics, be criminal offences? Is punishment for certain criminal behaviours right if they are strongly influenced by an individual's genes? If germ line gene editing can rid families of Huntington's disease, should they be free to use it? Can cloning an adult human ever be acceptable? And if tackling climate change means seeding the oceans with billions of genetically engineered algae, should it be done?

These are but a handful of the increasingly urgent and often intensely personal questions that our advancing understanding of life pushes us to ask. The only way to find acceptable answers is through constant, honest and open debate. Scientists have a special role to play in these discussions because it is they who must explain clearly the benefits, risks and dangers of each step forward. But it is society as a whole that must take the lead in the discussions. Political leaders must be fully engaged with these issues. Too few of them today take sufficient notice of the huge impact science and technology have on our lives and economies.

But the time for politics is *after* the science not *before*. The world has seen too often how things can go horribly wrong when the reverse is true. During the Cold War, the Soviet Union was able to build a nuclear bomb and send the first human into space. But work on genetics and crop improvements were severely damaged, because, for ideological reasons, Stalin backed the charlatan Lysenko who rejected Mendelian genetics. People starved as a consequence. More recently, we have witnessed the delays in action brought about by climate change deniers, who have ignored or actively undermined scientific understanding. Debates about the public good need to be driven by knowledge, evidence and rational thinking, and not by ideology, unsubstantiated beliefs, greed or political extremes.

But make no mistake, the value of science itself is not up for debate. The world needs science and the advances it can offer. As self-aware, ingenious and curiosity-driven humans, we have a unique opportunity to use our understanding of life to change the world. It is up to us to do what we can to make life better. Not only for our families and local communities, but also for all the generations to come, and for the ecosystems that we are an inextricable part of. The living world around us not only provides us humans with an endless source of wonder, it also sustains our very existence.

WHAT IS LIFE?

This is a big question. The answer I got at school was something like the MRS GREN list, which states that living organisms exhibit Movement, Respiration, Sensitivity, Growth, Reproduction, Excretion and Nutrition. It is a neat summary of the sorts of things that living organisms *do*, but it is not a satisfying explanation of what life *is*. I want to take a different approach. Based on the steps we have taken to understand five of biology's great ideas, I will draw out a set of essential principles that we can use to define life. These principles will then allow us to get deeper insight into how life works, how it got started, and the nature of the relationships that bind together all life on our planet.

Of course, many others have attempted to answer this question. Erwin Schrödinger emphasized inheritance and information in his prescient 1944 book *What is Life?* He proposed a 'code script' for life, which we now know is written in DNA. But he ended his book by making a suggestion that almost borders on vitalism: he argued that to really explain how life works, we might need a new and as yet undiscovered type of physical law.

A few years later the radical British-Indian biologist J. B. S. Haldane wrote another book, also called *What is Life?*, in which he declared, 'I am not going to answer this question. In fact, I doubt if it will ever be possible to give a full answer.' He compared the feeling of being alive to the perception of colour, pain or effort, suggesting that 'we cannot describe them in terms of anything else'. I have sympathy for Haldane's view, but it does rather remind me of the US Supreme Court Justice Stewart, who, in 1964, defined pornography by saying, 'I know it when I see it.'

The Nobel prizewinning geneticist Hermann Muller was not so hesitant. In 1966 he offered a 'stripped down' definition of a living thing as simply 'that which possesses the ability to evolve'. Muller correctly identified Darwin's great idea of evolution by natural selection as core to thinking about what life is. It is a mechanism – in fact, the only mechanism we know of – that can generate diverse, organized, purposeful living entities without invoking a supernatural Creator.

The ability to evolve through natural selection is the first principle I will use to define life. As shown in the chapter on natural selection, it depends on three essential features. To evolve, living organisms must reproduce, they must have a hereditary system, and that hereditary system must exhibit variability. Any entity that has these features can and will evolve.

My second principle is that life forms are bounded, physical entities. They are separated from, but in communication with, their environments. This principle is derived from the idea of the cell, the simplest thing that clearly embodies all the signature characteristics of life. This principle invokes a physicality of life, which excludes computer programs and cultural entities from being considered as life forms, even though they can appear to evolve.

My third principle is that living entities are chemical, physical and informational machines. They construct their own metabolism and use it to maintain themselves, grow and reproduce. These living machines are co-ordinated and regulated by managing information, with the effect that living entities operate as purposeful wholes.

Together, these three principles define life. Any entity which operates according to all three of them can be deemed to be alive.

The extraordinary form of chemistry that underpins life needs more elaboration for a full appreciation of how living machines work. A central feature of that chemistry is that it is built around large polymer molecules, formed mainly from linked atoms of carbon. DNA is one of them and its core purpose is to act as a highly reliable long-term store of information. To this end, the DNA helix shields its critical information-containing elements – the nucleotide bases – at the core of the helix, where they are stable and well-protected. So much so that scientists who study ancient DNA have been able to sequence DNA obtained from organisms that lived and died a very long time ago, including DNA from a horse that had been frozen in permafrost for nearly a million years!

But the information stored in the DNA sequence of the genes cannot remain hidden and inert. It must be transformed into action, to generate the metabolic activities and physical structures that underpin life. The information held in chemically stable and rather uninteresting DNA needs to be translated into chemically active molecules: the proteins.

Proteins are also carbon-based polymers, but in contrast to DNA, most of the chemically variable parts of proteins are located on the *outside* of the polymer molecule. This means that they influence the three-dimensional shape of the protein and also interact with the world. This is ultimately what allows them to perform their many functions, building, maintaining and reproducing the chemical machine. And unlike DNA, if proteins are damaged or destroyed, the cell can simply replace them by building a new protein molecule.

I cannot imagine a more elegant solution: different configurations of linear carbon polymers generate both chemically stable information storage devices and highly diverse chemical activities. I find this aspect of life's chemistry both utterly simple and completely extraordinary. The way life couples complex polymer chemistry with linear information storage is such a compelling principle that I speculate that it is not only core to life on Earth, but is also likely to be critical for life wherever else it may be found in the universe.

Though we and all other known life forms depend on carbon polymers, we should not be limited in our thinking about life by our experience of life's chemistry on Earth. It is possible to imagine life elsewhere in the cosmos that uses carbon in different ways, or life that is not built on carbon at all. The British chemist and molecular biologist Graham Cairns-Smith proposed in the 1960s a primitive life form based on self-replicating particles of crystalline clay, for example.

Cairns-Smith's imagined clay particles were based on silicon, a popular choice of science fiction writers when they imagine otherworldly life forms. Like carbon, silicon atoms can make up to four chemical bonds and we know they can form polymers: these are the basis of silicon sealants, adhesives, lubricants and kitchenware. In principle, silicon polymers might be large and varied enough to contain biological information. However, despite silicon being far more abundant on Earth than carbon, life here is based on carbon. That might be because under the conditions found on the surface of our planet silicon does not form chemical bonds with other atoms as readily as carbon does, and it does not therefore produce enough chemical diversity for life. It would be foolish, though, to rule out the possibility that silicon-based life, or for that matter life based on other chemistries altogether, might thrive in different conditions found elsewhere in the universe.

When thinking about what life is, it is tempting to draw a sharp dividing line between life and non-life. Cells are clearly alive and all organisms made from collections of cells are alive too. But there are other life-like forms that have a more intermediate status.

Viruses are the prime example. They are chemical entities with a genome, some based on DNA, others on RNA, which contains genes needed to make the protein coat that encapsulates each virus. Viruses can evolve by natural selection, thus passing Muller's test, but beyond that things are less clear. In particular, viruses cannot, strictly speaking, reproduce themselves. Instead, the only way they can multiply is by infecting the cells of a living organism and hijacking the metabolism of the infected cells.

So when you catch a cold, viruses enter the cells that line your nose and use your nose cell's enzymes and raw materials to reproduce the virus many times. So many viruses are produced, in fact, that the infected cell in your nose ruptures, releasing thousands of cold viruses. These new viruses infect nearby cells and get into your bloodstream to infect cells elsewhere. It is a highly effective strategy for a virus to perpetuate itself, but means that the virus cannot operate separately from the cellular environment of its host. In other words, it is completely dependent on another living entity. You could almost say that viruses cycle between *being* alive, when chemically active and reproducing in host cells, and *not* being alive, when existing as chemically inert viruses outside a cell.

Some biologists conclude that their strict dependence on another living entity means that viruses are not truly alive. But it's important to remember that almost all other forms of life, including ourselves, are also dependent on other living beings.

Your familiar body is in fact an ecosystem made up of a mixture of human and non-human cells. Our own 30 trillion or so cells are outnumbered by the cells of diverse communities of bacteria, archaea, fungi and single-celled eukaryotes that live on us and inside us. Many people carry with them larger animals too, including a variety of intestinal worms and the tiny eight-legged mites that live on our skin and lay their eggs in our hair follicles. Many of these intimate, non-human companions depend heavily on our cells and bodies, but we also depend on some of them too. For example, bacteria in our guts produce certain amino acids or vitamins that our cells cannot make for themselves.

And we should not forget that every single mouthful of the food we eat is produced by other living organisms. Even many microbes, such as the yeast I study, are completely dependent upon molecules usually made by other living organisms. These include glucose and ammonia for example, that are needed for making carbon- and nitrogen-containing macromolecules.

Plants appear to be rather more independent. They can draw carbon dioxide out of the air and water from the earth, and use the energy of the sun to synthesize many of the more complex molecules they need, including carbon polymers. But even plants rely on bacteria found in or near their roots that capture nitrogen from the atmosphere. Without them they cannot make the macromolecules of life. In fact, this is something that, so far as we know, no eukaryote can do for itself. That means there is not a single known species of animal, plant or fungus that can generate its own cellular chemistry entirely from scratch.

So perhaps the most genuinely independent life forms – the only ones with some claim to be fully independent and 'free-living' – are some that might at first seem rather primitive. These include the microscopic cyanobacteria, often called blue-green algae, that can both photosynthesize and capture their own nitrogen, and the archaea that get all their energy and chemical raw materials from volcanically active hydrothermal vents deep below the sea. Strikingly, these relatively simple organisms have not only survived for far longer than we have, but they are also more self-reliant than we are.

The deep interdependency of different life forms is also reflected in the fundamental make-up of our cells. The mitochondria that produce the energy our bodies need were once entirely separate bacteria – ones that had mastered the ability to make ATP. Through some accident of fate that took place around 1.5 billion years ago, some of these bacteria took up residence inside another type of cell. Over time, the host cells became so dependent on the ATP made by their bacterial guests that the mitochondria became a permanent fixture. The cementing of this mutually beneficial relationship probably marked the beginning of the entire eukaryote lineage. With a reliable supply of energy, the cells of eukaryotes were able to become bigger and more complex. This, in turn, precipitated the evolution of today's exuberant diversity of animals, plants and fungi.

This all demonstrates that there is a graded spectrum of living organisms that ranges from wholly dependent viruses, through to the much more self-sufficient cyanobacteria, archaea and plants. I would argue that these different forms are all alive. That's because they are all self-directed physical entities that can evolve by natural selection, although they are also all dependent to varying degrees on other living organisms.

From this broader perspective on life grows a richer view of the living world. Life on Earth belongs to a single, vastly interconnected ecosystem, which incorporates all living organisms. This fundamental connectedness comes not only from their deep interdependency, but also from the fact that all life is genetically related through its shared evolutionary roots. This perspective of deep relatedness and interconnectedness has long been championed by ecologists. It has its origins in the thinking of the early nineteenth-century explorer and naturalist Alexander von Humboldt, who argued that all life is bound together by a holistic web of connections. Unexpected as it may be, this interconnectivity is core to life, and should give us good reason to pause and think more deeply about the impact human activity has on the rest of the living world.

The organisms that live on the many branches of life's shared family tree are astonishingly varied. But that variety is outshone by their far greater and more fundamental similarities. As chemical, physical and informational machines, the basic details of their operations are the same. For example, they use the same small molecule, ATP, as their energy currency; they rely on the same basic relationships between DNA, RNA and protein; and they use ribosomes to make their proteins. Francis Crick argued that the flow of information from DNA to RNA to protein was so fundamental to life that he called it the 'Central Dogma' of molecular biology. Some have since pointed out minor exceptions to the rule, but Crick's key point still stands.

These deep commonalities in life's chemical foundations point to a remarkable conclusion: life as it is on Earth today started *just once*. If different life forms had emerged several times independently, and had survived, it is extremely unlikely that their descendants would all conduct their basic operations in such a similar way.

If all life is part of the same vast family tree, what kind of seed did that tree grow from? Somehow, somewhere, a very long time ago inanimate and disordered chemicals arranged themselves into more ordered forms that could perpetuate themselves, copy themselves and eventually gain the all-important ability to evolve by natural selection. But how did this story, which is eventually our story too, actually start?

The Earth was formed a little over 4.5 billion years ago, at the dawn of our solar system. For the first half a billion years or so, the surface of the planet was too hot and unstable to have allowed the emergence of life as we know it. The oldest unambiguously identified fossil organisms yet found lived around 3.5 billion years ago. That gives a window of a few hundred million years for life to get up and running. That's a longer expanse of time than our minds can readily comprehend, but is rather a small fraction of the total history of life on Earth. For Francis Crick, it seemed too improbable that life could have started here on Earth in the time available. That's why he suggested that life must have emerged elsewhere in the universe and been delivered here in either a partially or fully formed state. But this rather evades, instead of answers, the crucial question of how life might have started from humbler beginnings. Today, we can give a credible, if presently unverifiable, account of that story.

The oldest fossils look rather similar to some of today's bacteria. This indicates that life may already have been quite well established by that point, with membrane-enclosed cells, a hereditary system based on DNA, and a metabolism based on proteins.

But which came first? Replicating DNA-based genes, protein-based metabolism, or enclosing membranes? In today's living organisms these systems form a mutually interdependent system that only works properly as a whole. DNA-based genes can only replicate themselves with the assistance of protein enzymes. But protein enzymes can only be built from the instructions held in the DNA. How can you have one without the other? Then there's the fact that both genes and metabolism rely on the cell's outer membrane to concentrate the necessary chemicals, capture energy and protect them from the environment. But we know that cells alive today use genes and enzymes to build their sophisticated membranes. It's hard to imagine how one of this crucial trinity of genes, proteins and membranes could have come about on its own: if you take one element away, the whole system rapidly comes apart.

The formation of membranes might be the easiest part to account for. We know that the kind of lipid molecules that make up membranes can form via spontaneously occurring chemical reactions

that involve substances and conditions thought
to have been present on the young Earth. And
when scientists put these lipids into water, they do
something unexpected: they assemble themselves
spontaneously into hollow, membrane-bounded
spheres that are about the same size and shape as
some bacterial cells.

With a plausible mechanism for forming
membrane-bounded entities, that leaves the question of whether DNA genes or proteins came first.
The best solution scientists have yet found for this
particular chicken and egg-type problem is to say
that neither of them did! Instead, it may have been
DNA's chemical cousin, RNA, that came first.

Like DNA, RNA molecules can store information. They can also be copied, with errors in
that copying process introducing variability. That
means RNA can act as a hereditary molecule
that can evolve. That's what RNA-based viruses
still do today. The other crucial property of RNA
molecules is that they can fold up to form more
complicated three-dimensional structures that can
function as enzymes. RNA-based enzymes are not

nearly as complex or versatile as protein enzymes, but they can catalyze certain chemical reactions. Several of the enzymes crucial to the function of today's ribosomes are made from RNA, for example. If these two properties of RNA were combined, they may have been able to produce RNA molecules that work as both gene and enzyme: a hereditary system and a primitive metabolism in the same package. What this would amount to is a self-sustaining, RNA-based living machine.

Some researchers think these RNA machines might have first formed within the rocks that surround deep-ocean hydrothermal vents. Tiny pores in the rock may have provided a protected environment, whilst the volcanic activity boiling out of the Earth's crust would have offered a steady flow of energy and chemical raw materials. In these circumstances, it's possible that the nucleotides needed to make RNA polymers could be made from scratch, by assembling them from simpler molecules. At first, metal atoms embedded in the rock may have acted as chemical catalysts, allowing reactions to proceed without the aid of biological enzymes. Eventually, after millennia of trial and error, this could have led to the formation of machines made from RNA that were alive, self-maintaining and self-replicating and that, sometime later, could have been incorporated into membrane-bounded entities. That would have been a landmark event in the emergence of life: the appearance of the first true cells.

The story I have just told you is plausible, but please remember it is also highly speculative. The first life forms left no trace, so it is very difficult to know what was happening at the dawn of life or even what precise state the Earth itself was in more than 3.5 billion years ago.

Once the first cells had successfully formed, however, it is easier to imagine what happened next. First, the single-celled microbes would have spread through the world, gradually, colonizing sea, land and air. Then, 2 billion years or so later, the larger and more complex – but for a very long time still single-celled – eukaryotes joined them. True multicellular eukaryotic organisms came much later, after another billion years or so had elapsed.

That means that multicellular life has been here for about 600 million years, just one sixth of life's total history. But in that time they've given rise to all the largest and most visible living forms that surround us, including towering forests, swarming colonies of ants, huge networks of underground fungi, herds of mammals on the African savannah, and very much more recently, modern humans.

All of this has happened through the blind and unguided, but highly creative, process of evolution by natural selection. But when considering life's successes, we should remember that evolutionary change can only happen efficiently when some members of a population fail to survive and reproduce. So although life as a whole has proved itself to be tenacious, long-lasting and highly adaptable, individual life forms tend to have a limited lifespan and a restricted ability to adapt when their environment changes. Which is where natural selection comes into play, killing off the old order and, if more suitable variants exist in a population, making way for the new. It seems that it is only through death that there can be life.

The ruthless winnowing process of natural selection has created many unexpected things. One of the most extraordinary of these is the human brain. So far as we know, no other living thing shares with us quite the same awareness of its own existence. Our self-conscious minds must have evolved, at least in part, to give us more leeway to adjust our behaviour when our worlds change. Unlike butterflies, and perhaps all other known organisms, we can deliberately choose and reflect upon the purposes that motivate us.

The brain is based on the same chemistry and physics as all other living systems. Yet somehow, from the same relatively simple molecules and well-understood forces, spring our abilities to think, to debate, to imagine, to create and to suffer. How all this emerges from the wet chemistry of our brains provides us with an extraordinarily challenging set of questions.

We know that our nervous systems are based on immensely complex interactions between billions of nerve cells (neurons) that make trillions of connections between themselves, called synapses.

Together, these unfathomably elaborate and constantly changing networks of interconnected neurons establish signalling pathways that transmit and process rich streams of electrical information.

As is so often the case in biology, we know most of this from studying simpler 'model' organisms, such as worms, flies and mice. We know quite a lot about the ways these nervous systems gather information from their environments through their sense organs. Researchers have done a thorough job of tracking the movement of visual, sound, touch, smell and taste signals through the nervous system, as well as mapping some of the neuronal connections that form memories, generate emotional responses, and create output behaviours, such as flexing muscles.

This is all important work, but it is only a beginning. We have barely scratched the surface of understanding how the interactions between billions of individual neurons can combine to generate abstract thought, self-consciousness, and our apparent free will. Finding satisfactory answers to these questions will probably occupy the twenty-first century and likely beyond. And I do not think we can rely only on the tools of the traditional natural sciences to get there. We will have to additionally embrace insights from psychology, philosophy and the humanities more generally. Computer science can help too. Today's most powerful 'AI' computer programs are built to mimic, in a highly simplified form, the way life's neural networks handle information.

These computer systems perform increasingly impressive data-crunching feats, but display nothing that even vaguely resembles abstract or imaginative thought, self-awareness, or consciousness. Even defining what we mean by these mental qualities is very difficult. Here, a novelist, a poet or an artist can help, by contributing to the basis of creative thoughts, by more clearly describing emotional states, or by interrogating what it really means to *be*. If we have more of a common language, or at least greater intellectual connection, between the humanities and the sciences to discuss these phenomena, we may be better placed to understand how and why evolution has allowed us to develop as chemical and informational systems that have somehow become aware of their own existence. It will take all our imagination and creativity to understand how imagination and creativity can come about.

The universe is unimaginably vast. By the laws of probability, it seems very unlikely that across all that time and space life – let alone sentient life – has only ever blossomed once, right here on Earth. Whether or not we will ever meet alien life forms is a different issue. But if we ever do, I am confident they, like us, will be self-sustaining chemical and physical machines, built around information-encoding polymers that have been produced through evolution by natural selection.

Our planet is the only corner of the universe where we know for certain life exists. The life that we are part of here on Earth is extraordinary. It constantly surprises us but, in spite of its bewildering diversity, scientists are making sense of it, and that understanding makes a fundamental contribution to our culture and our civilization. Our growing understanding of what life is has great potential to improve the lot of humankind. But this knowledge goes even further. Biology shows us that all the living organisms we know of are related and closely interacting. We are bound by a deep connectedness to all other life: to the crawling beetles, infecting bacteria, fermenting yeast, inquisitive mountain gorillas and flitting yellow butterflies that have accompanied us during our journey through this book, as well as to every other member of the biosphere. Together, all these species are life's great survivors, the latest descendants of a single, immeasurably vast family lineage that stretches back through an unbroken chain of cell divisions into the far reaches of deep time.

As far as we know, we humans are the only life forms who can see this deep connectivity and reflect on what it might all mean. That gives us a special responsibility for life on this planet, made up as it is by our relatives, some close, some more distant. We need to care about it, we need to care for it. And to do that we need to understand it.

ACKNOWLEDGEMENTS

David and Rosie Fickling, for all their efforts to make this book accessible; and to friends and colleagues in my lab and beyond over the years, for discussions and disagreements about the nature of life. Finally, to Ben Martynoga, for helping me greatly and making this book enjoyable to write.